Lecture Notes in Electrical Engineering

Volume 564

The book series *Lecture Notes in Electrical Engineering* (LNEE) publishes the latest developments in Electrical Engineering - quickly, informally and in high quality. While original research reported in proceedings and monographs has traditionally formed the core of LNEE, we also encourage authors to submit books devoted to supporting student education and professional training in the various fields and applications areas of electrical engineering. The series cover classical and emerging topics concerning:

- Communication Engineering, Information Theory and Networks
- Electronics Engineering and Microelectronics
- Signal, Image and Speech Processing
- Wireless and Mobile Communication
- Circuits and Systems
- Energy Systems, Power Electronics and Electrical Machines
- Electro-optical Engineering
- Instrumentation Engineering
- Avionics Engineering
- Control Systems
- Internet-of-Things and Cybersecurity
- Biomedical Devices, MEMS and NEMS

For general information about this book series, comments or suggestions, please contact leontina. dicecco@springer.com.

To submit a proposal or request further information, please contact the Publishing Editor in your country:

China

Jasmine Dou, Associate Editor (jasmine.dou@springer.com)

India

Swati Meherishi, Executive Editor (swati.meherishi@springer.com)
Aninda Bose, Senior Editor (aninda.bose@springer.com)

Japan

Takeyuki Yonezawa, Editorial Director (takeyuki.yonezawa@springer.com)

South Korea

Smith (Ahram) Chae, Editor (smith.chae@springer.com)

Southeast Asia

Ramesh Nath Premnath, Editor (ramesh.premnath@springer.com)

USA, Canada:

Michael Luby, Senior Editor (michael.luby@springer.com)

All other Countries:

Leontina Di Cecco, Senior Editor (leontina.dicecco@springer.com)
Christoph Baumann, Executive Editor (christoph.baumann@springer.com)

** Indexing: The books of this series are submitted to ISI Proceedings, EI-Compendex, SCOPUS, MetaPress, Web of Science and Springerlink **

More information about this series at http://www.springer.com/series/7818

K. Shankar · Mohamed Elhoseny

Secure Image Transmission in Wireless Sensor Network (WSN) Applications

 Springer

K. Shankar
School of Computing
Kalasalingam Academy of Research
and Education
Virudhunagar, Tamil Nadu, India

Mohamed Elhoseny
Faculty of Computers and Information
Mansoura University
Mansoura, Egypt

ISSN 1876-1100   ISSN 1876-1119   (electronic)
Lecture Notes in Electrical Engineering
ISBN 978-3-030-20818-9   ISBN 978-3-030-20816-5   (eBook)
https://doi.org/10.1007/978-3-030-20816-5

This Springer imprint is published by the registered company Springer Nature Switzerland AG
The registered company address is: Gewerbestrasse 11, 6330 Cham, Switzerland

Preface

Wireless sensor networks (WSN) is a hot research topic since it is being applied in diverse fields. The recent interesting applications of WSN include target tracking, source localization, traffic surveillance, and healthcare monitoring require vision capabilities. In such applications, multimedia data such as images, videos, audio tracks, and animal sounds need to be transmitted using special multimedia sensor devices with additional capabilities, i.e., camera. The increasing depth and volume of multimedia data make it a more rewarding target for cybercriminals and state-sponsored espionage or sabotage. At the same time, due to the deployment of sensors in inaccessible areas, the probability of occurrence of different types of attacks is very high. A study conducted by Juniper Research in 2018 predicted that cybercriminals may steal an estimated amount of 33 billion records by the year 2023. Hence, it is difficult to predict the exact nature of future threats in WSN and how to combat it.

Researchers are constantly exploring novel and secure methods to transmit data, especially multimedia information in WSN. Since sensors have limited processing power, limited storage, low bandwidth and energy, traditional security measures designed for resource-rich networks, for instance LAN, are not suitable for resource-constrained WSN. In the presence of such limitations, it becomes mandatory to devise lightweight security solutions for data transmission in WSNs. Due to the existing problems in public key cryptography such as expensive key computation, longer keys, vulnerable keys to brutal force attack, key distribution and maintenance, it is not preferred for WSN. Keeping these issues in mind, we aim to devise lightweight security solutions, especially for WSN.

The unique feature of this book is that the WSN is leveraged for the secure transmission of digital images (DIs) using lightweight cryptography (LWC) techniques. The book covers state-of-the-art DI security techniques such as encryption, watermarking, steganography, and data hiding model utilizing LWC method with the help of various swarm intelligence-based optimization algorithms. By performing an extensive set of experiments, we provide a detailed performance evaluation in comparison with existing methods. We discuss a wide range of techniques to guarantee high security level, efficient energy consumption, and reasonable image quality. With innovative and promising theoretical and experimental results presented in this book, users can make use of it to combat against potential threats in information security. With a wide range of objects tested under these processes, the outcomes showed that any kind of multimedia information can be secured through the techniques proposed. High security level, efficient energy consumption, and reasonable image quality are guaranteed in the processes discussed in this book. The book details the techniques that kindle the research interests among budding scientists who are provided with suggestions in terms of security level, energy consumption, and image quality.

Salient Features of the Book

- Provides a comprehensive overview of techniques and processes involved in digital image security systems
- Details the cutting-edge techniques in a step-by-step fashion for clear and precise understanding
- Robust and promising methods with proven results for secure transmission of multimedia information
- Designed, developed, and written for a wide range of audience whom include students, research scholars, professionals, and a book for lifetime

Virudhunagar, India   K. Shankar
Mansoura, Egypt   Mohamed Elhoseny

Contents

Chapter 1
Introduction

Abstract WSN has a variety of multimedia-based information like image, video and secret data transmission process. For this process, the quality and the security of sensor nodes are critical. This chapter examined the background and difficulties of image security in WSN and Lightweight Cryptography (LWC) methods. Lightweight encryption strategy envelops quicker encryption and by analyzing the computing time, it expands the general lifetime of the sensor network. The fundamental reason for LWC in WSN is that its unique communication has been mixed or enciphered whereas the outcome is known as the cipher content or cryptogram. It is incorporated into block; the stream ciphers along with hash function are made to deliver the sturdy security for WSN image transmission process. Besides improving the nature of the images and security, the LWC optimization techniques also resemble PSO, GWO, and CSA with steganography, information data hiding and watermarking models. Toward the end of this chapter, the author discussed the vital performance measures utilized to analyze the security level of images in the network system.

Keywords Wireless Sensor Network (WSN) · Digital image · Optimization · Light Weight Cryptography (LWC) · Application · Security and ciphers

1.1 Digital Image Security

Digital Images (DI) turned out to be progressively essential in daily life due to which security of such information is critical. The principle reason behind the security of images is to maintain a strategic distance from or disregard the secret data access by the unapproved user, through communication in WSN. Digital Image security has many applications to its credit, in specific multimedia application, internet-based work, military application and social media along with every online application [1]. These applications need to control the access to images and then give way to confirm the honesty of the users.

© Springer Nature Switzerland AG 2019
K. Shankar and M. Elhoseny, *Secure Image Transmission in Wireless Sensor Network (WSN) Applications*, Lecture Notes in Electrical Engineering 564,
https://doi.org/10.1007/978-3-030-20816-5_1

1.1.1 Characteristics of Security

Image encryption and decryption models offer protected transmission and capacity technique for image over web. In order to make sensor nodes monetarily suitable as well as due to its vitality confinements, computation and correspondence abilities are restricted. Practically, each security approach requires a specific measure of assets [2]. Another significant issue is the way that remote channels are innately untrustworthy. It has numerous qualities to improve the security level and among everything, three attributes [3] are essential which are discussed in the following section.

Confidentiality: It is the term used to prevent the exposure of data to unapproved people or frameworks. The verified individuals can decipher the image content whereas nobody else can access it. It guarantees that no one can comprehend the received message other than the person who has the decipher key [4].

Integrity: Image cannot be altered by a non-approved individual. The DI cannot be altered or changed and is considered abuse when an image is effectively adjusted during transmission. Data security frameworks normally ensures message integrity and improve image privacy [5].

Authenticity: It is imperative that before they are shared in social networking, such interpersonal communication sites avoid controlled substance sharing via web-based networking media. The image has to be sure that the precise individuals have appeared or it remains as an adjusted form that utilizes different image processing softwares [6].

Availability: WSN is prone to various accessibility attacks which may seriously reduce the system tasks. The network even avoid applicable regions of an observed field from being detected by any sensor node. In spite of the fact that it is anything but a guard for attacks against accessibility, cryptography has turned into core safeguard systems in various sorts of networks [7].

1.1.2 History of Digital Image Security in WSN

DI security assumes a significant job in guarding the data. In spite of the fact that few fruitful strategies exist for image security, they are still under investigation to support their execution. These systems are segregated into two classes, in particular, DI encryption and DI hiding with some different models [8]. When exchanging DI through Wireless Sensor Network (WSN), the sensor nodes might be sent to extensive and difficult-to-reach regions, where the remote channel may be controlled by unapproved individuals. Notwithstanding the intrinsic issues when attempting to guarantee the secrecy, the transmission flow may likewise be liable to uprightness assaults [9]. In addition, it is hard to apply JPEG-based methodologies straightforward to asset-restricted sensors due to high computational unpredictability. Thus one need to think about how to guarantee high-security level at least computational overhead for asset-constrained sensors. Some critical depictions of WSN security (Fig. 1.1) are clarified below.

- The essential safeguard component in WSN is Light Weight cryptography (LWC). In simple terms, LWC is a set of methods to change data into a lot of indistinguishable information. At that point, it must be perused by the recipient who has the relating secret key [10].

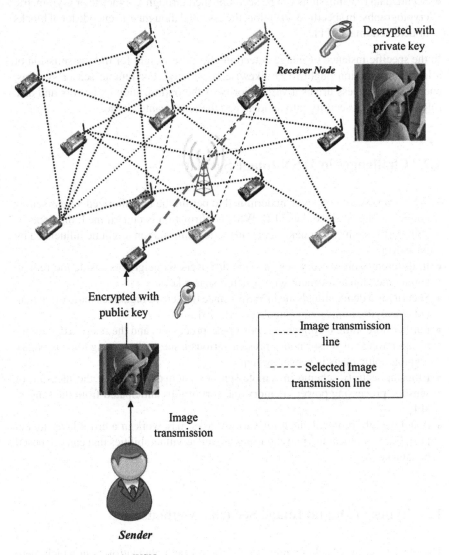

Fig. 1.1 Image security in WSN

- Enabling security for WSN is a fairly troublesome assignment. Besides, in WSN, information transmission requires big number of assets than the processing functions which possibly direct the selection of specific security systems.
- Thus the encryption instruments ought to be assessed by code range, information measure after encryption, processing time and power utilization.
- Since the security components might be unreasonably stringent for WSN, the manner in which the visual information gets detected and processed, might be abused to advance the security.
- Secure data transmissions can be accomplished through symmetric or asymmetric cryptography. In specific encryption, the essential thought is to encode lot of blocks of detected images [11].

In the specific instance of image detection, the extra weight for the transmission of a lot of information ought to be likewise considered. For remote sensor systems, energy effectiveness drives a vast majority of the optimization endeavors and as a rule it transforms security into an ideal and the most wanted issue.

1.2   Challenges in WSN Image Security

- WSN framework privacy is undermined to permit the unauthorized access sensor nodes and the detected data [12]. WSN confirmation is undermined; the dependability of system components and transmitted information might be influenced by false data.
- In different words, every sensor node discovers its neighbors inside the remote communication assortment with which it shares keys.
- Security is a demanding issue in WSNs since the sensor organizes more often than not conveying unforgiving situations [10, 13].
- Furthermore, little recollections, powerless processors and the restricted communication range of sensor nodes present various issues in executing the customary cryptographic plans in sensor systems.
- Efficient security scheme in remote sensor systems is made by the measure of sensors, processing power, memory and sort of jobs anticipated from the sensors [14].
- A lightweight protocol stack for various tasks may work in a cross-layer model [15]. Every protocol has particularities as well as vulnerabilities that can be abused by attackers.

1.3   Types of Digital Image Security Methods

The security against hacking attacks on web or accessible platforms in which there exist a distinctive information security system for multimedia information. Securing information with high-security condition is the top most priority nowadays. For most of the part, though high verify working conditions are performed and information is secured with encryption and decryption strategies or methods, yet that system utilizes just single encryption along with decryption keys [16]. For securing images in WSN transmission model, numerous procedures are available among which, three strategies are chosen i.e., Image encryption and hiding, and share creation [17]. From these procedures, the keys permit the validation of source nodes since they have the best possible keys. Moreover, the keys would be required to recoup the first information and confidentiality is additionally given.

1.3.1 Image Encryption

Image encryption is the best technique to shield the attackers from releasing the privacy of images. For security assurance, users normally encrypt the image before transmitting it to the server. Image encryption is taking their courses for great blend of speed, security and computational space [18]. These techniques can simply handle the DI features like size, repetition of information, strong relationships among contiguous pixels and so forth.

(a) Chaotic encryption

It was proposed by Matthews in 1989 that increases the research scholars' attention towards image encryption innovation based on chaotic systems. Most chaotic encryption algorithms utilize chaotic arrangement as the way to encrypt the information [19]. The key is delivered by logistic or different maps by making use of the chaotic value.

(b) Homomorphic encryption

It enables calculation to be completed on cipher content and at that point, the outcomes of the tasks performed can be decrypted by the data handler. At present, the data owner can get the same message from the off chance that it is performed on two plain images. A proficient encryption algorithm, that fulfills the homomorphic [20] property to perform encryption and decryption on medicinal images, is proposed. These images can be utilized to store in cloud and operations can be performed on it.

(c) Advanced Encryption Standard (AES) and Data Encryption Standard (DES)

These two encryption algorithms are primary to encipher the content and all things are considered in bumbling for the image encryption. Since, images have intrinsic features, for example, abundant repetition, there exists a strong relationship between among pixels. Along these lines, one can undoubtedly estimate the neighbors of a pixel in an image.

(d)   **Light Weight Cryptography (LWC)**

Lightweight Cryptography is one of the developing research regions in cryptography. The cryptographic algorithms are proposed for use in gadgets with low or very low assets. It is an area of established cryptographic algorithm that is relevant for asset-obliged gadgets in WSN. The encrypted information is just covered up at all huge bits. Through this methodology, the sender can keep the information inaccessible from unauthorized individuals and the receiver can safely access the sender's message [21].

1.3.2   Digital Steganography (DS)

Digital Steganography is a part of data hiding, a procedure that inserts message into cover contents and is utilized in numerous fields, for example, medical systems, law authorization, copyright insurance, access control and so on. It contains a high subtlety that diminishes the aggressor doubt of finding hidden message and is very hard to be identified by Stego image apparatuses. Any distortion if occurs to the cover information after implanting procedure, it builds interest among attackers. With expanding power of Steganography strategy in addition to cryptographic strategies, an extra layer of security is provided for the message at the time of communication. In this steganography-based security investigation, wavelet transformation was also used with LWC methods.

Pros of DS

- In any application, this could be utilized when one user sends an image to another user. It is utilized to shield personalities and profitable information from robbery, unapproved review, or potential harm by covering the message inside an unsuspicious image.
- It guarantees security, capacity, and vigor, the three required parts of steganography that make it helpful in hiding the trade of data through content archives and building up secret message.
- It predominantly serves to corporation governments and law authorization offices which can impart subtly.

1.3.3   Digital Watermarking

Watermark is a significant course of multimedia innovation inside the field of data security [22]. It ensures the copyright of the first information by installing secret data, for example, the watermark in the first information using different strategies. The chapter in this book favored Singular Value Decomposition (SVD) with optimization systems from this investigation that expand security, hide information as well as bring maximum robustness. The target of the watermarking model is to install water-marks in the first information and disconnect the digital watermark data from digital watermark information. It incorporates three vital procedures such as security, hidden information and maximum robustness. WSNs Digital Image insurance methodology models the watermarking to secure the detected information from attack models.

1.3.4 Data Hiding Model

In high-level data security that utilizes data hiding procedure, the installation limit, information security, great visual recuperation of hidden information and robust-ness are serious issues. The hidden data should be exchanged on safe transmission medium like WSN with the goal that nobody except the sender or collector, know the very presence of data [23]. Since the message is prone to mistakes, it does not get any consideration from unapproved users which protect the secret message. This evalua-tion of numerous strategies are utilized like Least-Significant-Bit (LSB), Pixel-Value Differencing (PVD) and Gray Level Modification (GVD) systems.

1.3.5 Secret Sharing Scheme

Secret sharing in different structures were winning before for different reasons more-over. Secrets were isolated into a number of shares and given to a similar number of individuals. Secret sharing varies and at the same time authorizes information protec-tion, accessibility and uprightness which no other security plot can accomplish. The secret image is reproduced with high-goals, i.e., practically lossless recuperation, by Lagrange's insertion when gathering any or more shadows. It is a fact that not exactly shadows recreate nothing of the secret image. Multiple secrets can be encrypted into numerous shares [24]. In any case, any two offers cannot recuperate two distinctive secret images and security of different offers utilizes XOR activity and cryptography method.

1.4 Lightweight Cryptography Mechanism

The requirement for LWC in WSN is vital to exchange the information, secret data or images from sender to recipient. The inspiration of LWC is to utilize less memory, less processing asset and less power supply to give security arrangement that can work over asset-restricted gadgets. The advancement of lightweight cipher began with the formation of an encryption technique. It is incorporated with the execution parameters such as controlled utilization, CPU significance and energy constraints. A lightweight secure information total strategy answer for dynamic node WSN which is dependent on a cryptographic methodology [25]. Key generation process includes complex scientific activities. In WSN conditions, these activities can be performed completely on 'encrypt and decrypt' model which is shown in Fig. 1.2.

For instance, the key length is decreased from 256 to 56 bits whereas the number of rounds that keep running in an encryption procedure, also get diminished from 48 to 16 and the method of design shifts from parallel to serial. Memory prerequisite is decreased from gigabytes to kilobytes. In this method, nodes set up the shared secret key and the open key with their neighbors. In this way, it gives encrypted information to the aggregator. LWC algorithms can be executed in equipment as programming. Equipment-based executions have elite properties yet have less adaptability and expense. From this, LWC offers more security than the prior ones and it is better to utilize symmetric ciphers in situations where confirmation and trustworthiness are of prime significance than non-revocation and privacy.

Challenges in LWC

- Conventional cryptography algorithm is not feasible for compelled gadgets due to its substantial key size, for example, RSA. RSA is not reasonable due to its expansive key size and high processing prerequisites.
- Third party programming downloaded from web on inserted devices is utilized in propelling the security assaults.
- Insecure frameworks, where the secret information is kept, is ensured that during message trade, the information holds its inventiveness and no adjustment is concealed by the framework.

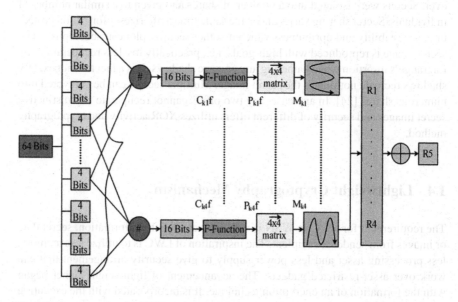

Fig. 1.2 Key generation for LWC

- WSN sensor nodes are itself considered as internet nodes that makes the validation procedure much critical. The trustworthiness of the information likewise becomes crucial and requires exceptional consideration towards holding its dependability.
- Efficient and secure correspondence between smart protests in the system is an essential goal. Smaller memory impressions and productivity makes LWC the most appropriate candidate for security answers by a number of brilliant articles.

1.4.1 Types of LWC

LWC has diverse methods and all these all systems provide the most extreme security in WSN during the information transmission process. A wide range of LWC makes the cipher more robust against any assault; the secret key is altered subsequent to encryption of each block of the pixels in an image.

Some vital types are 'Block cipher, hash function, stream cipher'. A detailed description of LWC is discussed in the following section and furthermore the overall characterization of the LWC is detailed in Fig. 1.3.

(1) **Block cipher**

The cipher accompanies the alternatives of 64, 96 and 128 bits key size. It is a symmetric key encryption technique that has encryption adjusts; each round depends on some scientific capacities to make confusion and diffusion. The increased number of rounds guarantees better security, yet in the long run, the outcomes will be incre-

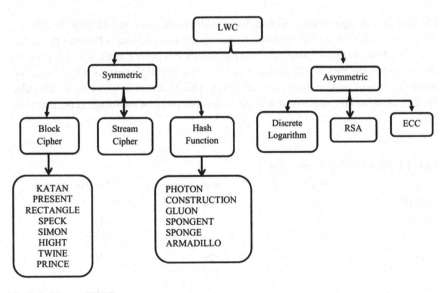

Fig. 1.3 Types of LWC

ment in the utilization of obliged energy. A few block ciphers including AES and some important highly secured block ciphers are discussed here. By and large, the block cipher algorithm can be separated into key scheduling capacity and encryption/decryption work. In order to assess this rule, a usage that has two functions was fabricated [26].

HIGHT: A minimal round function, without the utilization of S-boxes, was taken and all tasks were basic calculations. It utilized 128-piece keys with 64-bit blocks through 32 rounds. The noticeable element of HIGHT was that it comprised of some operations, for example, XOR, addition mod 28, left bitwise turn and the detail of ciphers are shown in Table 1.1 [26].

PRESENT: In order to achieve productive utilization of space, the block cipher was executed as encryption alike earlier. This cipher has 31 rounds and its different variations range from 2520 to 3010 GE (entryway Equivalent) to give sufficient security levels.

KATAN: KATAN accomplishes a little impression of 802 GE. It is a quicker cipher and for security reasons, the key generation and the assault recuperates the full 80-bit key.

SIMON and SPECK: Both were proposed by their authors in June 2013 [26]. Both are nonlinear capacities and are less amiable to programming executions. 3-bit S-boxes can be minimal, however, the way that 3 do not partition its 32, 64, and 128 complicates their utilization.

RECTANGLE: This cipher in LWC permits lightweight equipment and quick programming usage. A cipher state can be envisioned as a 4×16 rectangular group of bits, which is the beginning stage in cipher.

(2) Hash function

The proposed image security approach considers hash function with three highlights and is satisfied for this examination i.e., pre-image obstruction, collision-opposition and second pre-image opposition. This hash work has been executed in 4-bit S-box, and it satisfies the PRESENT plan criteria. The two principle parts of a hash function are development and compression functions. This function is to create an irregular number mask and the mask is then XOR with other piece of the image gets encrypted [27].

Table 1.1 Specifications for block cipher

Ciphers	Key size (bits)	Block size (bits)	Number of rounds
HIGHT	128	64	32
PRESENT	128	64	31
KATAN	80	64	254
SIMON	256	128	72
SPECK	256	128	34
RECTANGLE	128	64	25

(3) **Stream cipher**

The stream cipher is instated obviously, where the primary output bits make the yield out of the work. Dispersions and changes are effective in both hardware and software. Key scheduling was planned in a stream cipher way. From the key schedules, the stream depends only on the key and asynchronous ones, where the keystream likewise relies upon the cipher image. It persistently access the secret key. Be that as it may, it outlines the requirement for elective thoughts that permit the decrease in the extent of internal state even in situations where the consistent key access cannot be performed [13]. The size of the cipher key is 256-piece key and 16-bit block estimate through 20 rounds in WSN security investigation.

Advantages of LWC

- These primitives vary from regular algorithms with the presumptions that lightweight natives are not planned for a wide range of applications and may be confined with the standard of the attacker.
- Stream ciphers are increasingly hard to actualize accurately, and inclined to short-comings depending on utilization. Since the standards are like a one-time cushion, the keystream has strict prerequisites.
- The chosen AE worldview is appropriate for WSN correspondence as no decryption is expected to confirm the integrity of the secret message.
- Asymmetric cryptography supports all security administrations and furthermore gives a protected component towards key sharing. The main constraint is that it has vast key size which makes the encryption at moderate speed and builds in a multifaceted nature.
- Block cipher dispersion process, for example, the mix-column layer utilized in AES, are likewise conceivable. Despite the fact that they have cryptographic favorable circumstances, they incur high equipment cost.
- It is basically many-to-one capacity since they map self-assertive length contributions to fixed length yields and the input is normally more than the output.
- Encryption that mixes both symmetric as well as asymmetric encryptions exploits the qualities of both the encryptions and limits their shortcoming.

Applications of LWC

- LWC is generally used as a piece of IoT development to increase model security with least memory and consume less power.
- Ciphers and hash capacity of LWC are proficient information to select the fitting algorithm and hardware/software for a specific application.
- The mass deployment of inescapable devices guarantees from one perspective with numerous advantages like ideal store network, military, budgetary, e-business application, etc.
- With far-reaching nearness of embedded computers in such situations, security is an endeavoring issue in light of the fact that the potential harm of malevolent assaults increases day by day.

- Also, essential utilization of block ciphers is examined as these are the building blocks for cryptographic hash capacities.
- Cryptographic keys are utilized for e-business applications, for example, digital money, copyright security and so on. Additionally, all LWCs (block cipher, stream ciphers and hash capacities) are connected to improve the security of IoT applications.

1.5 Optimization Techniques for WSN Security

In WSNs, different optimization procedures are utilized to improve the security and reduce the power consumption and vitality utilization of LWC encryption algorithms. Dynamic enhancements give greater adaptability by constantly advancing a WSN/sensor node at the time of runtime, giving a better adjustment to change application prerequisites. Here in this book, three swarm-based best enhancement models i.e., PSO, GWO and CSA are discussed.

(i) Particle Swarm Optimization (PSO)

It is a swarm insight meta-heuristic algorithm motivated by social behavior of animals, for example, bird flocks or fish shoals. Here, every molecule position is assessed as indicated by the fitness function. The best molecule is indicated by the particles' past best position to refresh the new position and speed of the PSO [28] technique. It is generally utilized in WSN sensor node transmission process that build up the molecule out of all swarm molecules.

(ii) Grey Wolf Optimizations (GWO)

GWO is inspired by chasing approach of grey wolves which are viewed as the best predators. Grey wolves by and large move in a pack of around 5–12. This enhancer really works alike the chasing conduct of grey wolves pack. The guidelines/orders made by alpha are to be obeyed by whatever is left of the pack/gathering. Henceforth, alpha has the to-most dimension in the progressive system. This GWO [29] is likewise the best optimal solution in WSN security model in comparison with past PSO algorithm.

(iii) Cuckoo Search Algorithm (CSA)

CS is an improvement algorithm developed by Yang and Deb [30]. The behavior is attractive from the lovely sounds created by cuckoos and their multiplication approach turns out to be aggressive in nature. These birds are alluded as brood parasites as it lay their eggs in public nests. One needs to take note of that each egg in a nest which represents an answer and a cuckoo egg represents another arrangement where the goal is to supplant the weaker solution by a new solution. From CSA [31] optimization, the ideal keys are improved and the most extreme throughput followed by security dimension of image transmission in WSN are compared with GWO and PSO algorithms.

1.6 Evaluation Metrics of Digital Image Security

Peak Signal to Noise Ratio (PSNR): Peak signal power to noise power is measured for image quality.

$$PSNR = \sum_{i=1}^{s} 20 \times \log_{10}\left(\frac{255^2}{MSE_i}\right) \tag{1.1}$$

Entropy: Image entropy shows the measure of data contained in an image. It may be picked as a measure of the detail given by an image. It is determined by the probability of the decrypted image.

$$Entr = \sum_{i=0}^{2N-1} P_i \log(1/L_i) \tag{1.2}$$

Throughput: The throughput is the quantity of image (bits) expressed through a time component (s).

Number of Pixel Exchange Rate (NPCR): It is utilized to check the impact created by one-pixel change upon the whole image. NPCR shows the level of various pixels between two images.

$$NPCR = \left(\sum D(i,j) \big/ R * C\right) * 100 \tag{1.3}$$

Correlation Coefficient (CC): Two adjacent pixels throughout the plain-image as well as ciphered image are executed in this process.

Here $M(a), M(b)$ are the mean values of the image

$$CC(a, b) = \frac{CON(a, b)}{\sqrt{M(a) * M(b)}} \tag{1.4}$$

Normalized Coefficient (NC): Measure of similarity of two images as a function of a time-lag is applied to one of them.

$$NC(a, b) = \sum_{i}^{R}\sum_{j}^{C} \frac{a_{(i,j)} * b_{(i,j)}}{\sum_{i,j}^{R,C}(a_{i,j})^2} \tag{1.5}$$

Hiding Capacity: The maximum hiding capacity is the maximum amount of data that can be hidden in an image.

Mean Absolute Error (MAE): It is the maximum absolute value i.e., the difference between the original image and the degraded image.

$$MSE = \frac{1}{N} \left(\sum_{j=1}^{N} (A_j - E_j) \right) \qquad (1.6)$$

Mean Square Error (MSE): MSE is a type of average or sum (or integral) of the square of the error between two images. Here A_{ij} and B_{ij} are actual and decrypted images.

$$MSE = \frac{1}{RC} \left(\sum_{i=1}^{R} \sum_{j=1}^{C} (A_{ij} - E_{ij}^2) \right) \qquad (1.7)$$

1.7 Conclusion

Security instruments might be imperative in WSN. It incorporates encryption, decryption and some other intermediate procedures. All the above are made available for WSN applications since the proposed LWC encryption model the images to guarantee the security with important imperatives. In this chapter, the outline of image security in WSN, its effects, benefits and utilization of LWC along with optimizations are contemplated and discussed. There are numerous approaches available to enable security and the primary one is cryptography strategy for sensor nodes which is essential to give fitting security administrations. By choosing this, LWC with optimization security method for sensor nodes is essential to give security benefits in WSNs. Open Key-based cryptographic plans are acquainted after getting rid of the disadvantages present in symmetric approaches. Finally, the performance measures utilized for image security i.e., PSNR, NC, CC and so on are examined. This chapter is exceptionally helpful for wireless multimedia sensor systems, conceivably supporting important research in the upcoming years.

References

1. Zhao, G., Yang, X., Zhou, B., & Wei, W. (2010, July). RSA-based digital image encryption algorithm in wireless sensor networks. In *2010 2nd International Conference on Signal Processing Systems* (Vol. 2, pp. V2-640–643). IEEE.
2. Bisht, N., Thomas, J., & Thanikaiselvan, V. (2016, October). Implementation of security algorithm for wireless sensor networks over multimedia images. In *2016 International Conference on Communication and Electronics Systems (ICCES)* (pp. 1–6). IEEE.
3. Mahrous, A. M., Moustafa, Y. M., & El-Ela, M. A. A. (2018). Physical characteristics and perceived security in urban parks: Investigation in the Egyptian context. *Ain Shams Engineering Journal, 9*(4), 3055–3066.

4. Aminudin, N., Maseleno, A., Shankar, K., Hemalatha, S., Sathesh kumar, K., Fauzil, et al. (2018). Nur algorithm on data encryption and decryption. *International Journal of Engineering & Technology, 7*(2.26), 109–118.
5. Ilayaraja, M., Shankar, K., & Devika, G. (2017). A modified symmetric key cryptography method for secure data transmission. *International Journal of Pure and Applied Mathematics, 116*(10), 301–308.
6. Shankar, K., & Eswaran, P. (2016). An efficient image encryption technique based on optimized key generation in ECC using genetic algorithm. In *Advances in intelligent systems and computing* (Vol. 394, pp. 705–714). New York: Springer.
7. Shankar, K., Devika, G., & Ilayaraja, M. (2017). Secure and efficient multi-secret image sharing scheme based on Boolean operations and elliptic curve cryptography. *International Journal of Pure and Applied Mathematics, 116*(10), 293–300.
8. Darwish, A., Hassanien, A. E., Elhoseny, M., Sangaiah, A. K., & Muhammad, K. (2017). The impact of the hybrid platform of internet of things and cloud computing on healthcare systems: opportunities, challenges, and open problems. *Journal of Ambient Intelligence and Humanized Computing,* 1–16. https://doi.org/10.1007/s12652-017-0659-1.
9. Lee, S., Jeong, S., Chung, Y., Cho, H., & Pan, S. B. (2011, May). Secure and energy-efficient image transmission for wireless sensor networks. In *2011 IEEE Ninth International Symposium on Parallel and Distributed Processing with Applications Workshops* (pp. 137–140). IEEE.
10. Rekha, R. N., & PrasadBabu, M. S. (2012). On some security issues in pervasive computing-light weight cryptography. *International Journal on Computer Science and Engineering, 4*(2), 267.
11. Sathesh Kumar, K., Shanka, K., Ilayaraja, M., & Rajesh, M. (2018). Sensitive data security in cloud computing aid of different encryption techniques. *Journal of Advanced Research in Dynamical and Control Systems, 9,* 2888–2899.
12. Gupta, D., Khanna, A., Shankar, K., Furtado, V., & Rodrigues, J. J. (2018). Efficient artificial fish swarm based clustering approach on mobility aware energy-efficient for MANET. *Transactions on Emerging Telecommunications Technologies,* 1–10. https://doi.org/10.1002/ett.3524.
13. Bokhari, M. U., & Hassan, S. (2018). A comparative study on lightweight cryptography. In *Cyber Security: Proceedings of CSI 2015* (pp. 69–79). Singapore: Springer.
14. Mary, I. R. P., Eswaran, P., & Shankar, K. (2018). Multi secret image sharing scheme based on DNA cryptography with XOR. *International Journal of Pure and Applied Mathematics, 118*(7), 393–398.
15. Manifavas, C., Hatzivasilis, G., Fysarakis, K., & Rantos, K. (2013). Lightweight cryptography for embedded systems—A comparative analysis. In *Data privacy management and autonomous spontaneous security* (pp. 333–349). Berlin, Heidelberg: Springer.
16. Sehrawat, D., & Gill, N. S. (2018). Lightweight block ciphers for IoT based applications: A review. *Journal of Applied Engineering Research, 13*(5), 2258–2270. ISSN 0973-4562.
17. Elhoseny, M., Yuan, X., El-Minir, H. K., & Riad, A. M. (2016). An energy efficient encryption method for secure dynamic WSN. *Security and Communication Networks, 9*(13), 2024–2031.
18. Elhoseny, M., Elminir, H., Riad, A., & Yuan, X. (2016). A secure data routing schema for WSN using elliptic curve cryptography and homomorphic encryption. *Journal of King Saud University—Computer and Information Sciences, 28*(3), 262–275.
19. Wang, X. Y., & Gu, S. X. (2014). New chaotic encryption algorithm based on chaotic sequence and plain text. *IET Information Security, 8*(3), 213–216.
20. Shankar, K., & Lakshmanaprabu, S. K. (2018). Optimal key based homomorphic encryption for color image security aid of ant lion optimization algorithm. *International Journal of Engineering & Technology, 7*(1.9), 22–27.
21. Shehab, A., Elhoseny, M., Muhammad, K., Sangaiah, A. K., Yang, P., Huang, H., et al. (2018). Secure and robust fragile watermarking scheme for medical images. *IEEE Access, 6,* 10269–10278. https://doi.org/10.1109/access.2018.2799240.
22. Ping, N. L., Ee, K. B., & Wei, G. C. (2007). A study of digital watermarking on medical image. In *World congress on medical physics and biomedical engineering 2006* (pp. 2264–2267). Berlin, Heidelberg: Springer.

23. Elhoseny, M., Shankar, K., Lakshmanaprabu, S. K., Maseleno, A., & Arunkumar, N. (2018). Hybrid optimization with cryptography encryption for medical image security in internet of things. In *Neural computing and applications* (pp. 1–15).

24. Shankar, K., Elhoseny, M., Kumar, R. S., Lakshmanaprabu, S. K., & Yuan, X. (2018). Secret image sharing scheme with encrypted shadow images using optimal homomorphic encryption technique. *Journal of Ambient Intelligence and Humanized Computing*, 1–13.

25. Hatzivasilis, G., Fysarakis, K., Papaefstathiou, I., & Manifavas, C. (2018). A review of lightweight block ciphers. *Journal of Cryptographic Engineering, 8*(2), 141–184.

26. Hong, D., Sung, J., Hong, S., Lim, J., Lee, S., Koo, B. S., et al. (2006, October). HIGHT: A new block cipher suitable for low-resource device. In *International Workshop on Cryptographic Hardware and Embedded Systems* (pp. 46–59). Berlin, Heidelberg: Springer.

27. Arfan, M. (2016, October). Mobile cloud computing security using cryptographic hash function algorithm. In *2016 3rd International Conference on Information Technology, Computer, and Electrical Engineering (ICITACEE)* (pp. 1–5). IEEE.

28. Singh, N., & Singh, S. B. (2017). Hybrid algorithm of particle swarm optimization and grey wolf optimizer for improving convergence performance. *Journal of Applied Mathematics, 2017.*

29. Shankar, K., & Eswaran, P. (2015). Sharing a secret image with encapsulated shares in visual cryptography. *Procedia Computer Science, 70*, 462–468.

30. Yang, X. S., & Deb, S. (2014). Cuckoo search: recent advances and applications. *Neural Computing and Applications, 24*(1), 169–174.

31. Shankar, K., & Eswaran, P. (2016). RGB-based secure share creation in visual cryptography using optimal elliptic curve cryptography technique. *Journal of Circuits, Systems and Computers, 25*(11), 1650138.

Chapter 2
An Optimal Light Weight Cryptography—SIMON Block Cipher for Secure Image Transmission in Wireless Sensor Networks

Abstract With rapid growth of multimedia applications, secure image transmission over Wireless Sensor Network (WSN) is a challenging task. So, image encryption techniques are used to meet the demand for real-time image security over wireless networks. In the proposed research, the security of Digital Images (DI) in wireless sensor network is enhanced using Light Weight Ciphers (LWC) which encrypts the input image through encryption process. To improve DI privacy and confidentiality, an innovative security model is proposed i.e. Lightweight SIMON block cipher. The proposed LWC encrypted the image along with optimal key selection and enhanced the image security level in the cloud. For key optimization, a meta-heuristic algorithm called Opposition-based Particle Swarm Optimization (OPSO) algorithm was presented. The proposed SIMON-OPSO achieved the minimum time in generating key value to decrypt the image. The simulation result demonstrated that the SIMON-OPSO algorithm improves the accuracy of DI security for all input images (Lena, Barbara, Baboon, and House) compared to existing algorithms.

Keywords Digital image security · WSN · Encryption · LWC · SIMON block cipher · Key optimization · OPSO algorithm

2.1 Introduction

WSN is a class of adhoc networks where quality constrained sensor node are sent for some sort of monitoring or control function. Recently, many works have proposed inventive solutions for upgrade the performance of those networks, exhibiting promising commitments. Numerous wireless sensor applications in digital images will have security prerequisites. Security of digital images is progressively essential since the digital communication in WSN [1], over open systems, happens as often as possible [2]. The 'security of images' is an application layer innovated to have a look at transmitted data against undesirable exposure so as to prevent the data from adjustment while in travel [3]. Our security methodology in WSN guarantees security, genuineness [4], trustworthiness and non-denial and conceal the protection from delicate regions, distinctive cryptographic and RoI extraction strategies, sym-

© Springer Nature Switzerland AG 2019

K. Shankar and M. Elhoseny, *Secure Image Transmission in Wireless Sensor Network (WSN) Applications*, Lecture Notes in Electrical Engineering 564, https://doi.org/10.1007/978-3-030-20816-5_2

metric as well as asymmetric encryption systems [5]. Wireless sensor networks are defenseless against a few kinds of assaults, which may bargain one or more of the security prerequisites recently depicted. Along these lines, giving security to WSN [6] is a fairly difficult assignment. Moreover, in WSN, image transmission requests a bigger number of assets than handling capacities, conceivably directing the adoption of specific security systems [7]. Consequently, encryption instruments ought to be assessed by code measure, data size after encryption, preparing time and power utilization. Encryption is one of the essential ways to ensure the security of sensitive data. Encryption algorithm performs different substitutions, brings changes on the plaintext and changes it into ciphered image [8]. In case of encryption strategies like DES, AES, the former strategy generates 64-bits whereas the latter strategy generate 128, 192, 256 bits. The purpose of public key encryption is to explain the key dispersion of encryption and decryption model [9].

The proposed LWC encryption algorithm is utilized for sight and sound applications, for example, image. Furthermore different measurable tests were performed so as to guarantee the performance of the algorithm in securing the information [10]. In the proposed LW, selection of a key in cryptography is essential since the security behind encryption [11] algorithm depends primarily on it. The quality of the encryption algorithm depends on the secrecy and length of the key, the initialization vector, and how all these cooperate [12]. A combination of optimization in LWC in the protected key exchange instrument secures the information in a proficient way to verify the image stored in the cloud [13]. Hence, a cipher of high key and plain text affectability are attractive and computationally fast with regards to the nature of encoded images [14].

This chapter is structured as follows. Section 2.1 is the introduction part whereas the Sect. 2.2 describes the existing image security models along with optimization methods. Section 2.3 emphasizes the need for LWC in image security. The DI security methodology is discussed under Sect. 2.4. Section 2.6 explains the results achieved through proposed DI security model and Sect. 2.7 conclude the current research work with future scope.

2.2 Literature Review

A new encryption method for image security i.e., Multiple key-based Homomorphic Encryption (MHE) procedure was proposed by Shankar et al. [15] in which an optimal key is chosen with the help of Adaptive Whale Optimization (AWO) algorithm. PSNR value was assumed as a fitness function for optimizing plain as well as cipher images. The first image was changed into blocks and afterwards adjusted using encryption process. This was accomplished with high security and the result was much superior to other encryption procedures.

A powerful stenography algorithm ought to have less embedding distortion and equipped to avoid visual and measurable location. So there is a degree to find the quality of data concealing system by considering the optimization issue proposed by

Praneeta and Pradeep [16]. This depicts the structure that can investigate the 'security quality' of data hiding strategy called 'versatile pixel pair matching system' and there are two parameters discovered such as punishment parameter and kernel parameter.

In 2018, Avudaiappan et al. [17] have proposed 'r' a double encryption system which is used to scramble the medical images. At first, Blowfish encryption was applied and the afterwards signcryption algorithm was utilized to affirm the encryption technique. From that point onwards, the Opposition-based Flower Pollination (OFP) was used to redesign both private and public keys.

Wavelet-based secret image sharing method was proposed by Shankar et al. [18] with encoded shadow images utilizing the optimal Homomorphic Encryption (HE) system. The encoded shadow can be recuperated by just picking some subsets of these 'n' shadows that makes it apparent and stack over one another. To enhance the security of the shadows, each shadow is encoded and decoded utilizing HE strategy. With regards to the image quality issues, the new Opposition-based Harmony Search (OHS) algorithm was used to choose the optimal key.

According to the Chaotic (C-function) process, the security was investigated by Shankar et al. [19] like dispersion and perplexity. Adaptive Grasshopper Optimization (AGO) algorithm, along with PSNR and CC fitness function, was proposed to pick the ideal secret key as well as the public key of the framework among the arbitrary numbers. The purpose of choosing versatile process is to upgrade high-security examination of the proposed method in comparison with existing techniques. Finally, the proposed methodology results were contrasted with existing security techniques and artistic works which inferred that the proposed method as a high performing one.

Gaber et al. [20] had displayed a trust model and used to process a trust level for every node and the Bat Optimization Algorithm (BOA) was utilized to choose the cluster heads dependent on three parameters: residual energy, trust value and the quantity of neighbors. The results demonstrated that the proposed model was energy efficient. In expansion, the outcomes exhibited that the proposed model accomplished longer network lifetime.

In 2017, Elhoseny et al. [21] had displayed K-inclusion model dependent on Genetic Algorithm (GA) to expand a WSN lifetime. In the look for the ideal dynamic spread, distinctive factors, for example, targets' positions, the normal devoured vitality, a lot of trials were directed utilizing diverse K-coverage cases. Compared to some cutting edge strategies, the proposed model improved the WSN's performance in regards to the measure of the devoured energy, the network lifetime, and the expected time to switch between various spreads.

2.3 Image Security in WSN: LWC

There is a need for legitimate security of the image in wireless network to evade the unapproved individual's entry into the essential data. The importance of the image is that it covers more sight and sound data which needs assurance. LWC is a type of image security strategy that offers secure transmission as well as storage technique for images in web network. On account of image security, the systems utilized for encryption can be said as the protection device for secret data. The encryption process is the plain data that can be changed to ciphered or ensured data, and can be read only by decrypting it.

2.4   Security System Model in WSN

Security enhancement of Digital images in wireless network is progressively critical in current situation. For guaranteeing and upgrading the security of those images, optimal LWC was utilized i.e., SIMON LWC block cipher was used along with key optimization method. LWC is a completely-parameterized group of encryption algorithms in WSN, here AODV routing protocol used. So a unique decision of designing parameters were used such as the number of rounds, square size, data estimate, key length (private and public) and pixel size of the DI. These parameters were utilized to locate the optimal keys in better encryption and decryption model quality SIMON ciphers. In addition, the key optimization model is opposition-based swarm intelligence technique i.e., OPSO. With the help of this methodology shown in Fig. 2.1, the DI was encrypted and optimal private key was used to decrypt the images. The fundamental use of the methodology is to change pixel values which implies that higher the adjustment in pixels values, more effective the image encryption will

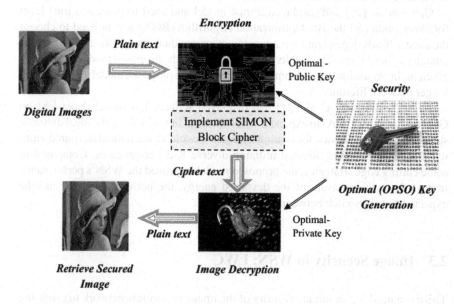

Fig. 2.1 Digital image security model in WSN: proposed

be which in turn enhance the encryption quality. A measure for encryption quality might be communicated as the deviation between the original and encrypted image.

2.5 Overview of Light Weight Ciphers for Security in Wireless Network

Normally, Light weight cryptographic algorithms are portrayed through block ciphers, hash capacities, verified encryption, and decryption modeling process. In this study, the block ciphers are only considered in DI security model in this network. The development of lightweight cipher started with the development of Advanced Encryption Standard (AES) strategy. Most lightweight ciphers are structure-dependent on AES and acquire a portion of the solid part in AES like S-box to avoid easy data interruption by hackers. In light of the structured parameters, just the ciphers were developed for cryptographic security analysis. Presently a number of ciphers are accessible, for instance, KATAN, KLEIN, SPECK, RECTANGLE, TWINE and SIMON and in this list of ciphers, SIMON ciphers were considered for the current study DI security.

2.5.1 SIMON Ciphers

A lightweight block cipher can be executed well since the block cipher is based on hash work and effective when connected in hardware. This cipher family comprises of 10 functions whereas two parameters differ in structuring the i.e., block and key size. For each block the key differs and these blocks depend on image pixels. The block measure changes in the middle of 32–128 bits, including the estimation of 16. It performs an action on fixed-size blocks of plain text and brings about a block of cipher content for each.

2.5.2 Characteristics of SIMON Cipher

SIMON cipher has direct and nonlinear attributes for security investigation in light of the block size and data. When utilizing a fixed input difference, one can think about a tree where every distinction at each round produces a few conceivable output differences.

- Assuming that the fundamental attributes can be stretched out to more rounds on the off chance, one can exploit the solid structure of the round capacity.
- The SIMON key calendars utilize the round consistency to kill the properties of key schedules and develop upon the quantity of pixel esteems in images.

- A single-key differential trademark and a single key differential for 15-round SIMON48 are present in lightweight block cipher [22].

The qualities present in this block cipher are not destined to be the best. The chosen optimizer yields the optimum solution whereas the quality found in this strategy is ensured to have the minimum number of active S-boxes.

2.5.3 Designing Model

At this point, when executing the cipher model for DI security in wireless network, a few conditions are viewed, for example, bit, round and encryption and decryption models. SIMON cipher with 2n-bit blocks are meant to the size as {16–64} and detail is clarified by the following equation

$$Enc\ DI = Cipher^i_{q_l}, \dots Cipher^n_{q_1}, \quad i > 1 \tag{2.1}$$

This Cipher $C^i_{q_l} \le i \le q$ are called 'round functions with round keys, this function are identical, its referred as an iterated block cipher.

Round configuration: In SIMON block cipher, the round function uses the inputs of 128-bit of plaintext and 128-bit key to produce 128-bit cipher message in 68 rounds. The operations used in SIMON encryption are depicted in n-bit words such as follows.

- Bitwise AND: This activity is performed on arbitrary two bits of n-bit words.
- Bitwise XOR: This activity is performed as an aftereffect of bitwise AND task and one of the bits from the lower block whereas the final value is XOR-ed with one of the arbitrary pieces from upper block that was at last XOR-ed with a key.
- Left bitwise revolution ROL, meant as $S^y(x)$ where y is the rotation count.

For encryption, SIMON round capacity is detailed as follows

$$RF(w_l, w_r, k_{round}) = \left((S^1(w_l)\ \&\ S^8(w_l)) \oplus S^2(w_l) \oplus w_r \oplus k_{round}, w_l\right) \tag{2.2}$$

Its inverse function is used to decrypt the image information and is described in Eq. (2.3),

$$RF^{-1}(w_l, w_r, k) = \left(w_r, (S^1(w_l)\ \&\ S^8(w_l)) \oplus S^2(w_r) \oplus w_l \oplus k_{round}\right) \tag{2.3}$$

The terms explained in Eq. (2.3) are, w_l is the left-most word of a given block, w_r is the right-most word and k_{round} is the appropriate round key. The diagrammatic representation of the round configuration is shown in Fig. 2.2.

Fig. 2.2 Round
configuration

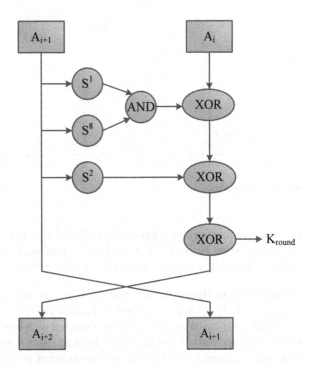

Key Generation: From the master key, SIMON cipher provides key expansion by producing all round keys. The selected SIMON64/128 configuration generates 44 32-bit sized round keys from the initial 128-bit master key. It does as such for a given round by consolidating the stored past round keys (where is the key words parameter) with consistent and a 1-bit round steady. The key expansion work uses the accompanying activities.

- Bitwise XOR, signified as $a \oplus b$.
- Right bitwise rotation ROR, meant as $s^{-c}(a)$ where c is the rotation count.

The function of key expansion can be understood from Eq. (2.4) and its diagrammatic representation is shown in Fig. 2.3.

$$Key_i(k, c, z_j) = F(k_{i+3}, k_{i+1}) \oplus S^{-1}(F(k_{i+3}, k_{i+1})) \oplus k_i \oplus c \oplus (z_j)_i \quad (2.4)$$

2.5.4 Optimal Key Selection: Opposition-Based Particle Swarm Optimization (OPSO)

Among a number of keys generated, the optimal one is chosen (it act as a private key for the end user) for decrypting the image. With the intention of choosing an optimal

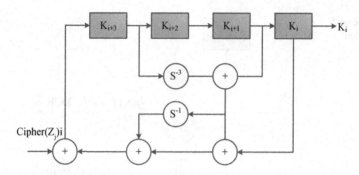

Fig. 2.3 Key generation

key, the optimization algorithm is selected which optimizes the value as minimum or maximum on the basis of objective of the current work. Here, the OPSO algorithm is used which will select the optimal key for decrypting the image.

Particle Swarm Optimization: Based on the swarm intelligence behavior, the researchers [23] developed an algorithm called PSO; it optimizes the key function. Especially, it is based on the research conducted with bird and fish flock movement behavior. Every particle that flies in the hunting space with a speed is adjusted by its own flying recognition and the flying knowledge of its friend in PSO.

Proposed OPSO Model: The performance of ordinary PSO algorithm was improved when an opposite solution is created. For every initialized particle, its corresponding opposite solution is initialized and the best solution is chosen by comparing these two (PSO and OPSO population). The opposite function can be calculated as Eq. (2.5).

$$\tilde{e}_j = f_j + g_j - e_j \qquad (2.5)$$

OPSO algorithm procedure

Step 1: Initialize the particles with random positions and their corresponding velocities. Here, the purpose of the proposed OPSO is to choose the optimal key value for decrypting the image. The initialized key function is performed as

$$Key(i) = Key_1, Key_2, \dots Key_n \qquad (2.6)$$

In Eq. (2.6), the term 'n' denotes the number of generated keys.

Step 2: The Fitness Function (FF) of the PSO algorithm is calculated on the basis of objective function of the research work.

$$FF = Opt\,(Key) \tag{2.7}$$

Calculate P$_{best}$ and G$_{best}$: At the starting point, the fitness value is generally determined for each and every particle. The optimal ones are elected as the G$_{best}$ and P$_{best}$ values among the fitness values. Consequent to that cycle, the current optimal fitness value is selected as the P$_{best}$ whereas the overall best fitness value is elected as the G$_{best}$. The fitness value of particle is compared with its P$_{best}$, if the current value is better then set P$_{best}$ equal to the current value.

Update the velocity and position: The equation used to update the velocity and position of the particles in the original PSO is as follows

$$v_i(t+1) = v_i(t) + d_1 rand(\,Pbest(t) - r_i(t)) + d_2 rand(\,Gbest - r_i(t)) \tag{2.8}$$

$$r_i(t+1) = r_i(t) + v_i(t+1) \tag{2.9}$$

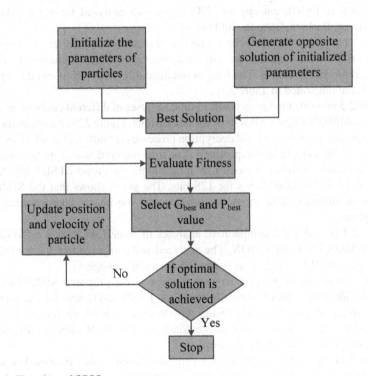

Fig. 2.4 Flowchart of OPSO

In Eqs. (2.8) and (2.9), V_i signifies the particle velocity, r_i signifies the current position of a particle, rand is a random number between (0, 1) and d_1, d_2 are learning factors, usually $d_1 = d_2 = 2$. According to the updated procedure on Eq. (2.8), the ith particle position is directed by the position of the global best solution and position best solution. Then the fitness for a new updated solution is identified.

Termination criteria: The new optimal value solution is then verified. If the desired optimum value is accomplished, then the OPSO procedure can be stopped; otherwise the process needs to be repeated from fitness evaluation. The flowchart of OPSO is illustrated in Fig. 2.4. Based on the optimal key selection (this is achieved by OPSO, the metaheuristic algorithm), the encrypted image is decrypted with high security in network.

2.6   Result and Analysis

The proposed digital image security is implemented in JDK 1.4 with 4 GB RAM and i5 processor. The result analysis section explains the efficiency of encryption and decryption processes in algorithm i.e. SIMON with OPSO. Further, the performance metrics such as PSNR, entropy and NPCR are also analyzed for the considered images (Lena, Barbara, Baboon and House).

Table 2.1 explains the results of digital image security analyses for the proposed SIMON-OPSO technique. For security analysis, the considered images are Lena, Barbara, Baboon and House. The images obtained after encryption and decryption processes are illustrated in Table 2.1.

Figure 2.5 presents the key generation time analyses of different encryption algorithms like SIMON-OPSO, SIMON-PSO and SIMON. Figure 2.5a, b explain the key generation time of encryption and decryption processes for different block sizes such as 32, 48, 64, 96 and 128 bits respectively. The time required to encrypt the image is high for maximum number of block size, for example, the proposed SIMON-OPSO took 16 s for 32 bits and 26.5 s for 128 bits. The graph shows that the SIMON-OPSO took minimum time to encrypt and decrypt the image when compared to other ciphers.

Table 2.2 provides the comparative analyses of three LWCs such as SIMON-OPSO, SIMON-PSO and SIMON. The analyzed and compared metrics are PSNR, entropy and NPCR for original as well as the encrypted image.

Figure 2.6 describes the performance metrics of the proposed SIMON-OPSO encryption algorithm. Figure 2.6a illustrates the PSNR (dB) value for the original and encrypted images. For Lena image, PSNR value was 63 dB for the original image and 60 dB for the encrypted image. Similarly, the PSNR value was examined for Barbara, Baboon and House image.

Figure 2.6b presents the entropy for original and encrypted images. For Lena image, the entropy value was 7.38 for the original image and 8 for the encrypted image. For Barbara image, entropy value was 7.98 for the original image and 7.96

Table 2.1 Image results for SIMON-OPSO technique

Image	Lena	Barbara	Baboon	House
Plain		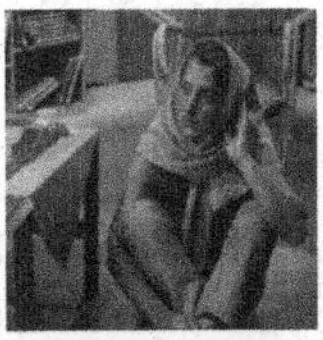	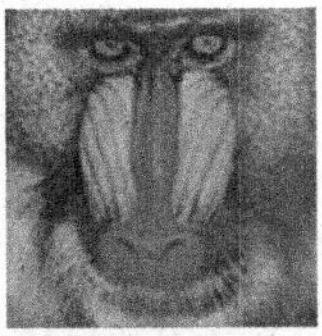	
Encrypted				
Decrypted			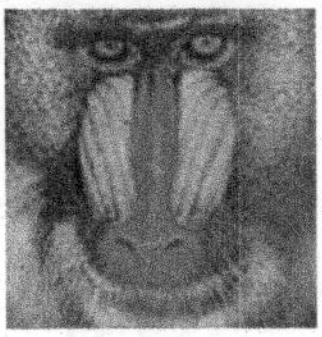	

for the encrypted image. For Baboon image, the highest entropy was achieved in the encrypted image. For house image, the entropies obtained for both original and encrypted images were same i.e. 8. Figure 2.6c explains the NPCR (%) for original and encrypted images (Lena, Barbara, Baboon and House images). The graph shows the NPCR value is high for the encrypted image. Hence, the security level of the proposed SIMON-OPSO is high compared to the plain text.

2.7 Conclusion

In this chapter, digital image security analysis in WSN by encryption technique is discussed along with key optimization. The images taken for security analysis were Lena, Barbara, Baboon and House. Here, the security level of digital images in WSN was enhanced by the implementation of LWCs during encryption and decryption processes. The proposed SIMON block cipher encrypted the image with desired PSNR, entropy and NPCR value and then decrypted the image on the basis of optimal key selection. Further, the presented SIMON-OPSO algorithm took minimum time to generate the optimal key when compared to other SIMON-PSO and SIMON encryption techniques in WSN. For future research on wireless sensor network image security, a comprehensive implementation of encryption algorithms need to be con-

Table 2.2 Comparative analysis

Metrics	Technique	Lena		Barbara		Baboon		House	
		Original	Encrypted	Original	Encrypted	Original	Encrypted	Original	Encrypted
PSNR (dB)	SIMON-OPSO	62.45	59.55	58.22	61.11	57.48	59.54	60.78	58.45
	SIMON-PSO	53.12	52.22	51.22	55.5	46.22	49.56	49.22	50
	SIMON	51.22	48.22	49.22	46.21	31.22	53.22	35.14	54.4
Entropy	SIMON-OPSO	7.33	7.99	7.95	7.89	7.69	7.96	7.99	7.99
	SIMON-PSO	6.99	7.89	7.89	7.85	7.02	7.89	6.85	7.86
	SIMON	6.89	7.69	7.78	7.86	7.66	7.66	7.82	7.82
NPCR (%)	SIMON-OPSO	82.22	96.45	93.15	94.45	89.55	96.22	92.122	89.22
	SIMON-PSO	79.42	81	86.22	86.45	76.22	92.18	81.22	79.22
	SIMON	72.22	79.45	69.45	88.5	82.14	84.55	76.54	76.2

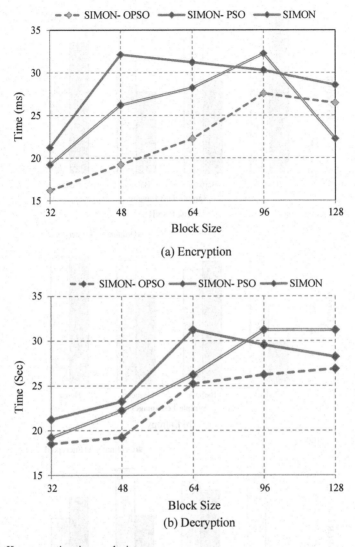

Fig. 2.5 Key generation time analysis

ducted with various metrics using innovative data encryption methods and hybrid optimization approaches. This can be applied for various purposes in cloud image security in the instance of different threats/attacks.

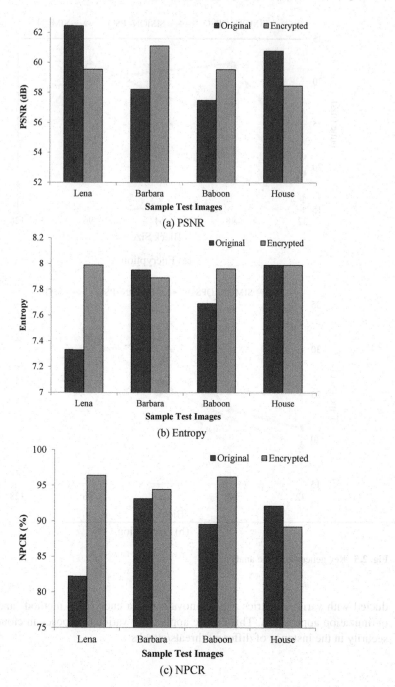

(a) PSNR

(b) Entropy

(c) NPCR

Fig. 2.6 Performance results for proposed security model

References

1. Riad, A. M., El-Minir, H. K., & El-hoseny, M. (2013). Secure routing in wireless sensor networks: A state of the art. *International Journal of Computer Applications, 67*(7).
2. Sathesh Kumar, K., Shankar, K., Ilayaraja, M., & Rajesh, M. (2017). Sensitive data security in cloud computing aid of different encryption techniques. *Journal of Advanced Research in Dynamical and Control Systems, 9*(18), 2888–2899.
3. Shankar, K. (2018). An optimal RSA encryption algorithm for secret images. *International Journal of Pure and Applied Mathematics, 118*(20), 2491–2500.
4. Elhoseny, M., Shankar, K., Lakshmanaprabu, S. K., Maseleno, A., & Arunkumar, N. (2018). Hybrid optimization with cryptography encryption for medical image security in Internet of Things. In *Neural computing and applications* (pp. 1–15).
5. Shankar, K., & Lakshmanaprabu, S. K. (2018). Optimal key based homomorphic encryption for color image security aid of ant lion optimization algorithm. *International Journal of Engineering & Technology, 7*(1.9), 22–27.
6. Elhoseny, M., & Hassanien A. E. (2019). Secure data transmission in WSN: An overview. In *Dynamic wireless sensor networks. Studies in systems, decision and control* (Vol. 165, pp 115–143). Cham: Springer.
7. Elhoseny M., & Hassanien A. E. (2019). Extending homogeneous WSN lifetime in dynamic environments using the clustering model. In *Dynamic wireless sensor networks. Studies in systems, decision and control* (Vol. 165, pp. 73–92). Cham: Springer.
8. Somaraj, S., & Hussain, M. A. (2015). Performance and security analysis for image encryption using key image. *Indian Journal of Science and Technology, 8*(35).
9. Karthikeyan, K., Sunder, R., Shankar, K., Lakshmanaprabu, S. K., Vijayakumar, V., Elhoseny, M., et al. (2018). Energy consumption analysis of virtual machine migration in cloud using hybrid swarm optimization (ABC–BA). *The Journal of Supercomputing*, 1–17.
10. Ramya Princess Mary, Eswaran, P., & Shankar, K. (2018). Multi secret image sharing scheme based on DNA cryptography with XOR. *International Journal of Pure and Applied Mathematics, 118*(7), 393–398.
11. Shankar, K., Devika, G., & Ilayaraja, M. (2017). Secure and efficient multi-secret image sharing scheme based on Boolean operations and elliptic curve cryptography. *International Journal of Pure and Applied Mathematics, 116*(10), 293–300.
12. Zhang, X., Seo, S. H., & Wang, C. (2018). A lightweight encryption method for privacy protection in surveillance videos. Algorithm. *Journal of Medical Systems, 42*(11), 208.
13. Usman, M., Ahmed, I., Aslam, M. I., Khan, S., & Shah, U. A. (2017). Sit: A lightweight encryption algorithm for secure internet of things. arXiv preprint arXiv:1704.08688.
14. Shankar, K., & Eswaran, P. (2017). RGB based multiple share creation in visual cryptography with aid of elliptic curve cryptography. *China Communications, 14*(2), 118–130.
15. Shankar, K., Lakshmanaprabu, S. K., Gupta, D., Khanna, A., & de Albuquerque, V. H. C. (2018). Adaptive optimal multi key based encryption for digital image security. *Concurrency and Computation: Practice and Experience*, 1–13.
16. Praneeta, G., & Pradeep, B. (2014). Security analysis of digital stegno images using genetic algorithm. In *Proceedings of the International Conference on Frontiers of Intelligent Computing: Theory and Applications (FICTA) 2013* (pp. 277–283). Cham: Springer.
17. Avudaiappan, T., Balasubramanian, R., Pandiyan, S. S., Saravanan, M., Lakshmanaprabu, S. K., & Shankar, K. (2018). Medical image security using dual encryption with oppositional based optimization algorithm. *Journal of Medical Systems, 42*(11), 208.
18. Shankar, K., Elhoseny, M., Kumar, R. S., Lakshmanaprabu, S. K., & Yuan, X. (2018). Secret image sharing scheme with encrypted shadow images using optimal homomorphic encryption technique. *Journal of Ambient Intelligence and Humanized Computing*, 1–13.
19. Shankar, K., Elhoseny, M., Chelvi, E. D., Lakshmanaprabu, S. K., & Wu, W. (2018). An efficient optimal key based chaos function for medical image security. *IEEE Access, 6,* 77145–77154.
20. Gaber, T., Abdelwahab, S., Elhoseny, M., & Hassanien, A. E. (2018). Trust-based secure clustering in WSN-based intelligent transportation systems. *Computer Networks, 146,* 151–158.

21. Elhoseny, M., Tharwat, A., Farouk, A., & Hassanien, A. E. (2017). K-coverage model based on genetic algorithm to extend WSN lifetime. *IEEE sensors letters, 1*(4), 1–4.
22. Bhoyar, P., Dhok, S. B., & Deshmukh, R. B. (2018). Hardware implementation of secure and lightweight Simeck32/64 cipher for IEEE 802.15.4 transceiver. *AEU—International Journal of Electronics and Communications, 90,* 147–154.
23. Dos Santos Júnior, J. G., & do Monte Lima, J. P. S. (2018). Particle swarm optimization for 3D object tracking in RGB-D images. *Computers & Graphics, 76.*

Chapter 3
An Optimal Lightweight RECTANGLE Block Cipher for Secure Image Transmission in Wireless Sensor Networks

Abstract The fast development of networking permits substantial documents, for example, multimedia images, to be effectively transmitted over the Wireless Sensor Networks (WSNs). An image encryption is generally used to guarantee the security as it may secure the images in the greater part. The process of securing the images in WSN against unauthorized users is a challenging one. For guaranteeing high security among Digital Images (DIs), Light Weight Cryptographic (LWC) algorithms are utilized which split the DI into a number of blocks; this will upgrade the level of security in WSN. Here, the proposed block cipher is RECTANGLE which separates the image in a bit-slice style; and improves the DI security level by encryption and decryption depending on the determination of optimal public key and private key individually. The key optimization was finished by the metaheuristic algorithm, for example, Opposition-based Grey Wolf Optimization (OGWO) which chosen optimal key on the basis of the most extreme PSNR value. The exhibited RECTANGLE–OGWO accomplished the least time to produce key which remained as an incentive to encrypt and decrypt the image. The result showed that the RECTANGLE–OGWO algorithm enhanced the accuracy of DI security for all the images (Lena, Barbara, Baboon, Airplane and House) when contrasted with existing algorithms.

Keywords Digital image (DI) security in WSN · Encryption · LWC · RECTANGLE block cipher · Grey wolf optimization

3.1 Introduction

The essential aspect of securing images in WSN is to ensure confidentiality, integrity, and authenticity. Diverse methods are available to secure image; in particular, encryption is one of the main processes to secure digital images in WSN [1]. To shield the DI from illegal assurance, information is encoded through cryptographic methods utilizing a secret key; this secret key can be communicated over WSN through secret [2] channel to the approved purchaser of the information. This strategy is exceptionally valuable for transmission of computerized image for its amazing encryption procedure and every information is in the encoded frame [3]. For most of the part,

© Springer Nature Switzerland AG 2019

K. Shankar and M. Elhoseny, *Secure Image Transmission in Wireless Sensor Network (WSN) Applications*, Lecture Notes in Electrical Engineering 564, https://doi.org/10.1007/978-3-030-20816-5_3

there are two principle keys to build the entropy; the variable secret key of the change procedure and the variable secret key of the LWC model [4]. Usually, cipher encryption algorithms are utilized for image encryption [5–7]. Block and stream ciphers are constructed with the help of a pseudo-random key sequence after which this arrangement is joined with the first content through elite or administrator [6]. Multi-stage applications are probably going to be the standard instead of special case for new applications of LWC [7]. The SIMON configuration can be summed up to SIMON-like ciphers in WSN [8], which utilize a similar structure and round function however both are extraordinary rotational constants [9]. The absence of structured method of reasoning and security assessment for SIMON and SPECK increased the attention among the cryptanalysts' who took a ton of examinations for a more thoughtful comprehension of this cipher [7]. So in this chapter, we propose the RECTANGLE block ciphers for the image security in WSNs. LWC underlines the effective usage of cryptographic algorithms and it is a relatively-young logical sub-field that is situated at the side of software engineering, electrical building, and cryptography [10].

The primary difference between the standard block ciphers and the lightweight block ciphers is the block size. As a rule, it is 32, 48 or 64 bits for a lightweight block cipher and equivalent to 64 or 128 bits for a customary block cipher [11]. A cipher is viewed as broken if the foe can decipher the secret key. On the off chance, if the assailant can as often as possibly decode the ciphered content without deciding the secret key, the cipher is said to be mostly broken [12]. Thus, barely it is incorporated with an IoT or cloud at the digital images. Thus, a lightweight and equipment-viable encryption approach is favored [10]. Besides, when individuals wish to replace images over an unreliable system, it ends up pivotal at that point to give a flat security. To be precise, over WSN an image requires protection from different security assaults [13]. This chapter has Sect. 3.2 under which recent literature is discussed followed by Sect. 3.3 investigating the background of LWC. The proposed procedure is discussed in Sect. 3.4 and Sect. 3.5 examines the outcomes with conclusion in Sect. 3.6.

3.2   Literature Survey

One of the objective functions of WSNs is detecting and gathering encompassing data intermittently through sensor nodes establishing the system and conglomerating to sink node for further processing of DI security in WSN Patil [14]. Mohamed et al. [15] played out an enlightening survey covering typical hierarchical WSN routing protocols utilizing a far reaching correlation dependent on their general exhibitions and application situations. In parallel to the improvement in image encryption mechanism, it is additionally all about the innovation advancement about image data theft. So in order to meet the developing innovation in data theft, one must look for a superior image encryption algorithm, as opined by Pan et al. [16]. In their study, the simulation tests were completed utilizing the traditional Lena image and with the existing pictures whereas the outcomes were segregated into histogram, pixel connection, data entropy, key space measure and key affectability. An epic procedure was proposed to enhance a chaos-based image encryption algorithm by Noshadian et al. [17]. The optimization technique yielded the parameters that led to the most minimal relationship among adjoining pixels or most noteworthy entropy. The double encryption methodology was used to encrypt the medical images in WSN; introduced by El-Shorbagy et al. [18].

At first, Blowfish Encryption was considered after which signcryption algorithm was used to affirm the encryption model. From that point onwards, the Opposition based Flower Pollination (OFP) was used to redesign the private and open keys. Avudaiappan et al. [19] assessed the execution of the proposed system by utilizing the performance metrics, for example, Peak Signal to Noise Ratio (PSNR), entropy, Mean Square Error (MSE), and Correlation Coefficient (CC). In 2018, Poonam et al. [20], in their work mentioned that a computerized watermark can be inserted as host information in spatial space as in recurrence area. In this work, a hybridized strategy consolidating Discrete Wavelet Transform (DWT) and Singular Value Decomposition (SVD) were proposed. In 2018, Shankar et al. [21] investigated the Chaotic (C—work) process in which the security was investigated like dispersion just as disarray. In view of the underlying conditions, distinctive random numbers were created for each array from disordered maps. Adaptive Grasshopper Optimization (AGO) algorithm, with PSNR and CC fitness exertion, was proposed to pick the ideal secret and open keys of the framework among the random numbers. The choice of versatile process is to upgrade high-security examination of the current proposed model contrasted with existing strategies [22].

3.3 Background of LWC in Image Security

The inspiration behind LWC is to utilize less memory, less processing asset, and less power supply to ensure that the security arrangement works even in case of asset-constrained gadgets in WSN. The LWC is generally more straightforward and quicker when compared with traditional cryptography methods. The current chapter explicitly examines the lightweight usage of symmetric-key block ciphers in equipment and programming structures. Among PRESENT, TEA, LED, KATAN, KLEIN, SPECK, RECTANGLE, TWINE and SIMON, the current chapter uses a RECTANGLE block cipher to secure the DI over WSN by bit-slice style.

3.4 Methodology: Digital Image Security in WSN

This research work examines the issue of digital image security in wireless networks and the challenge to upgrade the security level of an image by LWC security model. Normally, LWCs are a completely-parameterized group of encryption algorithms. So, a few planning parameters are used i.e., a defined number of rounds, block size, data estimate, key length (private and public) and pixel size of the DI. With the assistance of these parameters, one needs to locate the optimal keys for better encryption as well as decryption of DI in WSN utilizing RECTANGLE block cipher. For optimal key selection, opposition-based metaheuristic algorithm i.e., Opposition-based Grey Wolf Optimization (OGWO) is recommended. One of the favorable circumstances in the proposed philosophy is that the RECTANGLE accomplishes an extremely-aggressive programming speed among the current lightweight block ciphers because of its bit-slice style. Likewise, it provides great security-performance among all DIs over WSN. To accomplish better outcomes, recommend a multipath routing protocol AODV is used to transmit data or DI over wireless networks.

3.4.1   Image Collection

In the proposed DI security analysis, some test images i.e. Lena, Barbara, Baboon, Airplane and House were considered. These images are highly secure with the usage of RECTANGLE block LWC. The images used for encryption and decryption analyses are shown in Fig. 3.1.

3.4.2   Image Security: Implementation of LWC Algorithm

LWC algorithms are described by four functions to be specific such as block ciphers, hash functions, confirmed encryption and decryption. The fundamental thought behind the RECTANGLE structure is to permit lightweight and quick executions utilizing bit-slice methods. RECTANGLE [23] utilizes SP-network and the substitution layer comprises of 16 numbers of 4×4 S-encloses parallel. The permutation layer is made out of 3 revolutions. As discussed in this chapter, RECTANGLE offers extraordinary performance in both equipment and programming conditions, which ensure enough adaptability to various application situations.

(1)   **RECTANGLE block cipher**

RECTANGLE is an iterated block cipher. The block length is 64 bits and the key length is 80 or 128 bits. A 64-bit plaintext or a 64-bit middle of the result, or 64-bit

Fig. 3.1 Sample digital images for security analysis

cipher content is by and large called as a cipher state. A cipher state can be imagined as a 4×16 rectangular cluster of bits, which is the starting point of the cipher named RECTANGLE.

Designing process of RECTANGLE block cipher

Round function: RECTANGLE is a 25-round SP-network cipher. Each one of the 25 rounds comprises of three stages: After the last round, there is a last addroundkey.

- AddRoundkey
- SubColumn
- ShiftRow

AddRoundkey: An easy bitwise XOR of the round subkey to the intermediate state.

SubColumn: In a RECTANGLE block cipher, the S-boxes are parallel utilized to four bits in a similar section. The activity of SubColumn is delineated in Eq. (3.1).

$$
\begin{pmatrix} p_{0,15} \\ p_{1,15} \\ p_{2,15} \\ p_{3,15} \end{pmatrix} \cdots \begin{pmatrix} p_{0,3} \\ p_{1,3} \\ p_{2,3} \\ p_{3,3} \end{pmatrix} \begin{pmatrix} p_{0,2} \\ p_{1,2} \\ p_{2,2} \\ p_{3,2} \end{pmatrix} \begin{pmatrix} p_{0,1} \\ p_{1,1} \\ p_{2,1} \\ p_{3,1} \end{pmatrix} \begin{pmatrix} p_{0,0} \\ p_{1,0} \\ p_{2,0} \\ p_{3,0} \end{pmatrix}
$$
$$
\downarrow s \quad \cdots \quad \downarrow s \quad \downarrow s \quad \downarrow s \quad \downarrow s \tag{3.1}
$$
$$
\begin{pmatrix} q_{0,3} \\ q_{1,3} \\ q_{2,3} \\ q_{3,3} \end{pmatrix} \cdots \begin{pmatrix} q_{0,3} \\ q_{1,3} \\ q_{2,3} \\ q_{3,3} \end{pmatrix} \begin{pmatrix} q_{0,3} \\ q_{1,3} \\ q_{2,3} \\ q_{3,3} \end{pmatrix} \begin{pmatrix} q_{0,3} \\ q_{1,3} \\ q_{2,3} \\ q_{3,3} \end{pmatrix} \begin{pmatrix} q_{0,3} \\ q_{1,3} \\ q_{2,3} \\ q_{3,3} \end{pmatrix}
$$

ShiftRow: A left rotation to each column over various balances i.e., Column 0 is not turned, row 1 is left pivoted more than 1 bit, row 2 is left turned more than 12 bits, and line 3 is left turned more than 13 bits. Let $\ll x$ mean left pivot over x bits, the activity ShiftRow is shown in the condition (3.2).

$$
\left(p_{0,15} \cdots p_{0,3} \ p_{0,2} \ p_{0,1} \ p_{0,0} \right) \xrightarrow{\ll 0} \left(p_{0,15} \cdots p_{0,3} \ p_{0,2} \ p_{0,1} \ p_{0,0} \right)
$$
$$
\left(p_{1,15} \cdots p_{1,3} \ p_{1,2} \ p_{1,1} \ p_{1,0} \right) \xrightarrow{\ll 1} \left(p_{1,14} \cdots p_{1,2} \ p_{1,1} \ p_{1,0} \ p_{1,15} \right)
$$
$$
\left(p_{2,15} \cdots p_{2,3} \ p_{2,2} \ p_{2,1} \ p_{2,0} \right) \xrightarrow{\ll 11} \left(p_{2,4} \cdots p_{2,8} \ p_{2,7} \ p_{2,6} \ p_{2,5} \right) \tag{3.2}
$$

Key Schedule: In a RECTANGLE block cipher, the keys are scheduled as either 80 or 128 bits. For an 80-bit seed key (user-provided key), the key is initially stored in an 80-bit key register and arranged as a 5×16 array of bits. From these varieties of bits, an ideal key is selected to encrypt (ideal public key) and decode (ideal private key) the digital image whereas the algorithm is introduced and named as OGWO.

3.4.3   Proposed Key Selection Model

By using OGWO, the optimal keys for both encryption and decryption processes can be selected. In the research work, the optimal keys are identified by image PSNR value.

Grey Wolf Optimization (GWO): GWO is a recently-presented developmental algorithm according to which the grey wolves have an effective generation more than chasing in pack. Two grey wolves (male and female) hold higher position and deal with alternate scalawags. It is like different metaheuristics, and in GWO [24, 25] algorithm, the inquiry starts by a populace of haphazardly-created wolves (candidate solutions). Figure 3.2 underneath illustrates the social hierarchy of wolves.

Opposition process: The execution of conventional GWO algorithm is improved by the generation of inverse arrangement. For each introduced parameter, its relating inverse arrangement is instated and the best arrangement is chosen by looking at these two (GWO and OGWO populace).

$$\tilde{x}_j = y_j + z_j - x_j \tag{3.3}$$

Steps involved in OGWO algorithm

The number of OGWO population is initialized. The function of OGWO is to select the optimal key value for encrypting as well as decrypting the image. The initialized key function is expressed as

$$Key(i) = \{K_1, K_2 \ldots K_{80}\} \tag{3.4}$$

The optimal key is chosen by the maximum PSNR value of DI. This is set to be the fitness function of OGWO algorithm. The fitness function is described in Eq. (3.5).

$$F_i = Max\_PSNR \tag{3.5}$$

To locate the separate solution (threshold) based on the fitness value, let the first best fitness solution be α, the second best fitness solution be β and the third best fitness solution be δ.

Updating

Here, it can be expected that the alpha (best applicant arrangement), beta and delta possess enhanced information about the potential area of the prey so as to scientifically recreate the chasing conduct of the grey wolves. Thus, the initial three best arrangements accomplished until now are accumulated and the other pursuit specialists (counting the omegas) are required to change their situations as per the situation of the best inquiry operator. For redundancy, the new solution $c(t + 1)$ is evaluated by utilizing the formulae referenced underneath.

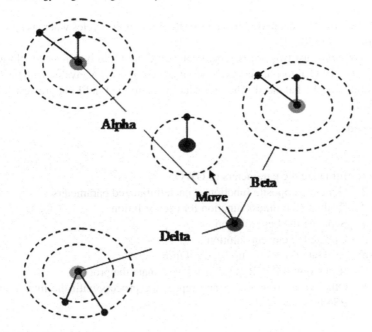

Fig. 3.2 Social hierarchy of wolves

$$D^{\alpha} = |C_1 \cdot c_{\alpha} - c|, \quad D^{\beta} = |C_2 \cdot c_{\beta} - c|, \quad D^{\delta} = |C_3 \cdot c_{\delta} - c| \qquad (3.6)$$

$$c_1 = c_{\alpha} - A_1 \cdot (D_{\alpha}), \quad c_2 = c_{\beta} - A_2 \cdot (D_{\beta}), \quad c_3 = c_{\delta} - A_3 \cdot (D_{\delta}) \qquad (3.7)$$

To have hyper-circles with various irregular radii, the subjective parameters A and C help the candidate arrangements. Examination and use are guaranteed by the versatile estimations of A and a. The versatile estimations of the parameters permit the GWO to travel it effectively between the examination and the use. With reducing A, half of the emphases is given to the examination ($|A| < 1$) and the other half is committed to the utilization. Encasing the direct, the ensuing conditions are used remembering the ultimate objective to give the numerical model.

$$D = |C \cdot c_P(t) - c(t)| \qquad (3.8)$$

The coefficient vectors are found by

$$A = 2a \cdot r_1 - a, \quad C = 2 \cdot r_2 \qquad (3.9)$$

where t demonstrates the current iteration, A and C are coefficient vectors, c_P is the position vector of the prey c which shows the position vector of a grey wolf. The parts of a are linearly decreased from 2 to 0 throughout the cycles whereas r_1, r_2 are random vectors in [0, 1].

The GWO has just two primary parameters (A and C) to be balanced. Notwithstanding, the GWO algorithm is kept as straightforward as conceivable with the least administrators to be balanced. The procedure is continued until the greatest PSNR is attained.

Pseudo code

Step 1: Initialize the parameters of GWO
Step 2: Generate opposite function of each initialized parameters
Step 3: Evaluate the fitness function for each solution
Step 4: Separate the best solution
Step 5: Update the current solution
Step 6: Evaluate fitness for updated solution
Step 7: If maximum PSNR is achieved, terminate the process
Step 8: Otherwise, go to step 3 and repeat the process until the maximum PSNR is achieved.

3.5 Implementation Results Analysis

The results for the proposed optimal LWC-block cipher for DI security were analyzed in this area. It was actualized utilizing the working stage of MATLAB 2016a with the system configuration, i5 processors with 4 GB RAM. Here the parameters considered were PSNR, Entropy, and MAE for proposed and conventional optimization algorithms.

A novel encryption technique creates the optimal Key sensitivity. Table 3.1 demonstrates all DIs with ciphered and decrypted images with histogram of those cipher images. In the security model, the image was segregated into bigger number of blocks which improved the execution. The entropy was expanded as the quantity of blocks got expanded. For the decryption procedure, ideal private key was only utilized which was in hexadecimal frame and length of key was 128 bits.

Figures 3.3 and 3.4 demonstrate the aftereffects of the proposed security strategy RECTANGLE with optimal key selection (OGWO) along that PSNR, Entropy and Mean absolute error (MAE) measurements. The convergence of the optimization appears in Fig. 3.3. If the iteration differs with PSNR, it likewise fluctuates in light of the optimal keys. Here the least PSNR was 21.25 dB and the greatest was 61.25 dB in OGWO model. When compared with GWO, a considerable PSNR was accomplished. At that point, Fig. 3.4 depicts the measures for various images such as Lena, Barbara, Baboon, Airplane and House with Original and ciphered images. The PSNR and entropy rate were the most extreme in ciphered images contrasted with unique images.

Table 3.1 Image results for the proposed security model

Image	Plain image	Histogram for plain image	Cipher image	Histogram for cipher image	Decrypted with optimal key
Lena					
Barbara					
Baboon					
House					
Airplane					

For example, in Lena, the PSNR value was 52.22 for plain and it was 58.22 for ciphered images and comparatively, the encrypted value for both were 7.58 and 7.89. The entropy relies upon the algorithm's multifaceted nature as well as the key and the images have certain effect. Figure 3.4c illustrates the MAE measures for plain and cipher images. The procedures of decryption need errorless input and output images to provide security for the information that they process.

The relative examination of the security model is shown in Table 3.2. Here the impressive comparison strategies are RECTANGLE, RECTANGLE-GWO and RECTANGLE-OGWO for all images. The PSNR rates of baboon were 39.22, 52.22 and 58.22 dB for plain images when compared to the PSNR values of ciphered images being 39.22, 43.22, 58.22 dB. Further the image results were analyzed in the table above. The minimum entropy was 6.45 in RECTANGLE plain images and the most extreme esteem was 7.99 in cipher images in the proposed model. In addition, MAE

Fig. 3.3 Iteration versus fitness

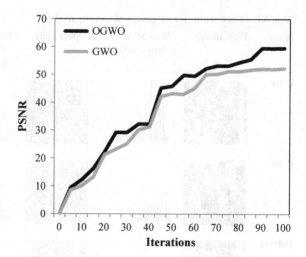

likewise provides better aftereffects in the proposed security model when compared with different systems.

Encryption, decryption and execution time of the proposed strategy is illustrated in Fig. 3.5. It can be observed that the proposed strategy consumed less time to encrypt/decrypt an image. For Lena image, the execution, encryption and decryption times were 0.03, 0.08, and 0.12 s. Accordingly, the time taken for analysis of images was determined

Table 3.3 shows the images without attack and with attack. The PSNR rate of salt and pepper attack of house image was 30.14 dB, pixel exchange was 29.22 and without attack the value was the most extreme i.e., 49.22 dB. The entropy estimations of the encrypted images were near the ideal value, which implies that the proposed encryption algorithm is exceptionally powerful against entropy attack. It is likewise the same for MAE in all attacks applied and without attack images.

3.6   Conclusion

The DI security analysis over WSN against unauthorized user access over the network was presented in this chapter. The analysis was performed by applying LWC block cipher and key optimization. An innovative cryptographic algorithm i.e. RECTAN-GLE block cipher was proposed to enhance the security level of DIs in WSN. The proposed RECTANGLE cipher performed well when compared to other existing block ciphers. With the help of RECTANGLE block cipher, digital images were encrypted and decrypted based on the optimal key selection. Furthermore, the presented OGWO algorithm took minimum time to generate an optimal key when compared to other SIMON-OPSO and SIMON encryption techniques. On further research work regard-

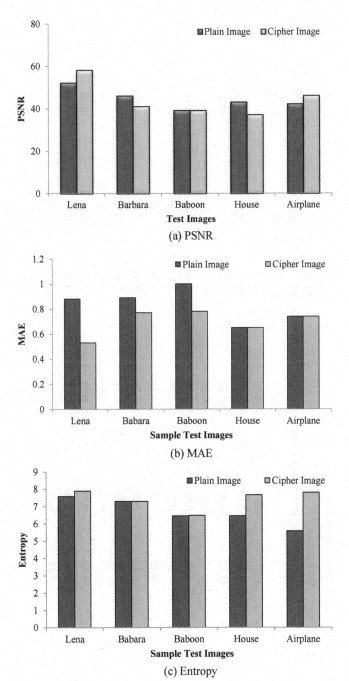

Fig. 3.4 Security measures analysis

Table 3.2 Comparative analysis for digital images

Images	Measures	Plain images				Cipher image			
		RECTANGLE	RECTANGLE-GWO	RECTANLE-OGWO		RECTANGLE	RECTANGLE-GWO	RECTANLE-OGWO	
Lena	PSNR	52.22	56.52	61.25		58.22	50.18	60.08	
Barbara		46.12	49.79	59.56		41.12	51.22	56.22	
Baboon		39.22	52.22	58.22		39.22	43.22	58.22	
House		43.12	56.45	61.22		37.22	40.18	49.22	
Airplane		42.22	49.55	59.52		46.22	52.22	56.22	
Lena	Entropy	7.58	6.99	7.05		7.89	7.92	7.99	
Barbara		7.29	6.78	7.06		7.29	7.79	7.82	
Baboon		6.45	6.78	6.89		6.48	7.58	7.96	
House		6.45	7.01	7.05		7.66	7.59	7.92	
Airplane		5.58	6.45	7.12		7.81	7.56	7.95	
Lena	MAE	0.88	0.57	0.65		0.53	0.42	0.38	
Barbara		0.89	0.59	0.58		0.77	0.59	0.28	
Baboon		1	0.58	0.52		0.78	0.51	0.29	
House		0.65	0.55	0.42		0.65	0.51	0.49	
Airplane		0.74	0.79	0.51		0.74	0.45	0.38	

Table 3.3 Comparison between attack and without attack

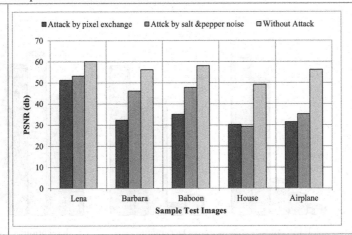

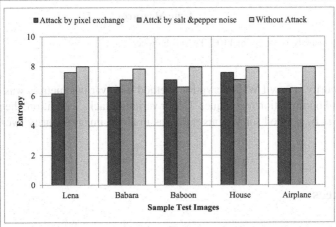

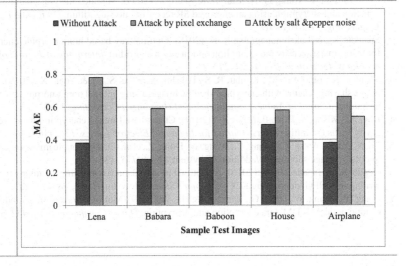

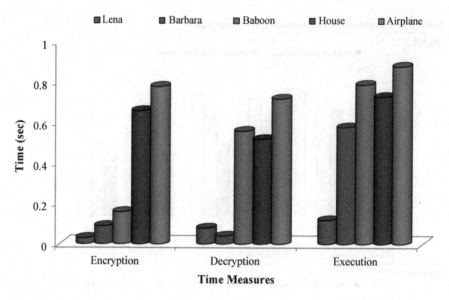

Fig. 3.5 Time analysis for optimal RECTANGLE cipher

ing DI security, a new LWC encryption method with various metrics was analyzed with hybrid optimization techniques. Under various attacks' condition, the proposed strategy can be applied for various purposes in cloud image security over wireless networks.

References

1. Li, X., Wang, Y., Wang, Q. H., Liu, Y., & Zhou, X. (2019). Modified integral imaging reconstruction and encryption using an improved SR reconstruction algorithm. *Optics and Lasers in Engineering, 112,* 162–169.
2. Shankar, K., & Lakshmanaprabu, S. K. (2018). Optimal key based homomorphic encryption for color image security aid of ant lion optimization algorithm. *International Journal of Engineering & Technology, 7*(1.9), 22–27.
3. Shankar, K., Elhoseny, M., Kumar, R. S., Lakshmanaprabu, S. K., & Yuan, X. (2018). Secret image sharing scheme with encrypted shadow images using optimal homomorphic encryption technique. *Journal of Ambient Intelligence and Humanized Computing,* 1–13.
4. Roy, A., Misra, A. P., & Banerjee, S. (2019). Chaos-based image encryption using vertical-cavity surface-emitting lasers. *Journal for Light and Electron Optics, 176,* 119–131.
5. Yu, N., Xi, S., Wang, X., Zhang, C., Wang, W., Dong, Z., et al. (2019). Double images encryption in optical image subtraction/addition 4F system. *Optik, 178,* 135–141.
6. Hegde, R., & Jagadeesha, S. (2016). An optimal modified matrix encoding technique for secret writing in MPEG video using ECC. *Computer Standards & Interfaces, 48,* 173–182.
7. Wang, Y., Zhang, W., Chen, G., Yang, X., & Hu, W. (2019). Multi-Gbit/s real-time modems for chaotic optical OFDM data encryption and decryption. *Optics Communications, 432,* 39–43.

8. Boukerche, A., & Sun, P. (2018). Connectivity and coverage based protocols for wireless sensor networks. *Ad Hoc Networks, 80,* 54–69.

9. Elhoseny, M., Shankar, K., Lakshmanaprabu, S. K., Maseleno, A., & Arunkumar, N. (2018). Hybrid optimization with cryptography encryption for medical image security in Internet of Things. In *Neural computing and applications* (pp. 1–15).

10. Elhoseny, M., Elminir, H., Riad, A., & Yuan, X. (2016). A secure data routing schema for WSN using elliptic curve cryptography and homomorphic encryption. *Journal of King Saud University—Computer and Information Sciences, 28*(3), 262–275.

11. Wang, X. Y., & Li, Z. M. (2019). A color image encryption algorithm based on Hopfield chaotic neural network. *Optics and Lasers in Engineering, 115,* 107–118.

12. Su, Y., Wo, Y., & Han, G. (2019). Reversible cellular automata image encryption for similarity search. *Signal Processing: Image Communication, 72,* 134–147.

13. Shankar, K., & Eswaran, P. (2017). RGB based multiple share creation in visual cryptography with aid of elliptic curve cryptography. *China Communications, 14*(2), 118–130.

14. Patil, B. S. (2015). Image security in wireless sensor networks using wavelet coding. *International Journal on Emerging Technologies, 6*(2), 239.

15. Mohamed, R. E., Ghanem, W. R., Khalil, A. T., Elhoseny, M., Sajjad, M., & Mohamed, M. A. (2018). Energy efficient collaborative proactive routing protocol for wireless sensor network. *Computer Networks, 142,* 154–167.

16. Pan, H., Lei, Y., & Jian, C. (2018). Research on digital image encryption algorithm based on double logistic chaotic map. *EURASIP Journal on Image and Video Processing, 2018*(1), 1–10.

17. Noshadian, S., Ebrahimzade, A., & Kazemitabar, S. J. (2018). Optimizing chaos based image encryption. *Multimedia Tools and Applications,* 1–22.

18. El-Shorbagy, M. A., Elhoseny, M., Hassanien, A. E., & Ahmed, S. H. (2018). A novel PSO algorithm for dynamic wireless sensor network multiobjective optimization problem. *Transactions on Emerging Telecommunications Technologies,* 1–14.

19. Avudaiappan, T., Balasubramanian, R., Pandiyan, S. S., Saravanan, M., Lakshmanaprabu, S.K., & Shankar, K. (2018). Medical image security using dual encryption with oppositional based optimization algorithm. *Journal of Medical Systems, 42*(11), 208.

20. Poonam, & Arora, S. M. (2018). A DWT-SVD based robust digital watermarking for digital images. *Procedia Computer Science, 132,* 1441–1448.

21. Shankar, K., Elhoseny, M., Chelvi, E. D., Lakshmanaprabu, S. K., & Wu, W. (2018). An efficient optimal key based chaos function for medical image security. *IEEE Access, 6,* 77145–77154.

22. Shankar, K., Lakshmanaprabu, S. K., Gupta, D., Khanna, A., & de Albuquerque, V. H. C. Adaptive optimal multi key based encryption for digital image security. *Concurrency and Computation: Practice and Experience,* 1–11.

23. Zhang, W., Bao, Z., Lin, D., Rijmen, V., Yang, B., & Verbauwhede, I. (2015). RECTANGLE: A bit-slice lightweight block cipher suitable for multiple platforms. *Science China Information Sciences, 58*(12), 1–15.

24. Teng, Z. J., Lv, J. L., & Guo, L. W. (2018). An improved hybrid grey wolf optimization algorithm. In *Soft computing* (pp. 1–15).

25. Shankar, K., & Eswaran, P. (2015). A secure visual secret share (VSS) creation scheme in visual cryptography using elliptic curve cryptography with optimization technique. *Australian Journal of Basic and Applied Sciences, 9*(36), 150–163.

Chapter 4
An Optimal Lightweight Cryptographic Hash Function for Secure Image Transmission in Wireless Sensor Networks

Abstract In the recent years, numerous security schemes have been proposed to secure the data and Digital Images (DI) over WSNs. Especially, encryption and decryption algorithms are structured and actualized to provide secrecy and security in WSN during the transmission of image-based information just as in storage. In this chapter, Lightweight Cryptography (LWC) based hash function is used for image security in WSN. The hash function keeps up different guidelines which contain a set of tenets with user details, IP address, public and private keys. The hash value of encryption was developed upon the optimal secret key and it was recognized by the Enhanced Cuckoo Search (ECS) optimization. In this ECS model, cuckoo birds choose the nests of various birds to leave its eggs i.e., optimal keys. Further impressive fitness function parameters such as Peak Signal to Noise Ratio (PSNR) were kept consistent in this research. The proposed system provided expanded security and adequately utilized the algorithm when compared with ordinary encryption and optimization strategies.

Keywords Encryption · Decryption · Security · WSN · Hash function · Cuckoo search optimization · Optimal key

4.1 Introduction

A Wireless Sensor Network (WSN) is a group of tiny power-constrained nodes that cover wide range of applications in the image security [1]. As of late, extra security and protection issues increased due to the fast advancements in the field of communication systems in WSN [2]. The need for privacy and standardized methods for communication with images and videos have turned out to be amazingly fundamental and also other related issues that ought to be mulled over WSN [3]. Image encryption systems attempt to change the original image to another image which is difficult to comprehend; to keep the secrecy of the image between clients [4]. In other words, it is basic that no one could become acquainted with the substance without a key for decryption [5]. In secret key cryptography, single key is utilized. Secret key cryptography incorporates DES, AES, 3DES, IDEA, Blowfish algorithms and

© Springer Nature Switzerland AG 2019

K. Shankar and M. Elhoseny, *Secure Image Transmission in Wireless Sensor Network (WSN) Applications*, Lecture Notes in Electrical Engineering 564,
https://doi.org/10.1007/978-3-030-20816-5_4

so on whereas public key cryptography incorporates RSA, Digital Signature along with Message Digest algorithms [6]. The chosen cryptographical model in the current research requires more time to specifically encode the image information [7]. The other issue is that the decrypted image should be equivalent to the first image. The principle points of interest in LWC hash function encryption approach include high adaptability in the encryption [8] framework design, accessibility of tremendous varieties of disordered frameworks, extensive, perplexing and various conceivable encryption keys with less complex structure [9]. This guarantees the strong encryption without trading off the ease of use framework as far as speed and robustness are concerned.

- Cryptography-based encryption is determined; it decides scientific conditions that manage its behavior [10].
- They are unusual, non-straight, and responsive to introductory conditions and even when there is an extremely slight change in the beginning stage, it can prompt noteworthy and diverse results.
- Computer data when transmitted must be accessed by the authorized party and not by any other individuals and the determination of key in cryptography is imperative since the security of encryption algorithm depends specifically on it [11].

In order to enhance the WSN security of LWC model, a number of optimizations are utilized such as Genetic Algorithm (GA), Particle Swarm Optimization (PSO) and so on. [12]. Due to the probability of success in an attack and accordingly the optimization of overall security of the network, the current research points out the necessity to model this problem as an optimization problem [12]. The hash computation uses the padded data along with functions, constants, word logical and algebraic operations, to iteratively generate a series of hash values [13]. The proposed work uses a hash function-LWC to find out the locations for storing secret image in WSN [14]. The secret image is encrypted using ECS with a hash function. This makes it even more secure and it works for color images too.

4.2 Literature Review

WSNs have attracted serious interest from both academia and industry because of their wide application in common and military situations. As of late, there is an incredible interest related to routing procedure in WSNs. Security perspectives in routing protocols have not been given enough consideration was discussed by Riad et al. in 2016 [15], since a large portion of the in WSNs have not been structured in view of security necessities. Shih et al. in 2013 [16] was proposed as an image based recovery algorithm to enhance the lifetime and security of a WSN. Wavelet-based secret image sharing scheme was proposed by Shankar et al. [17] with encrypted shadow images that utilizes optimal Homomorphic Encryption (HE) procedure. The scrambled shadow can be recouped by picking some subsets of these 'n' shadows that makes straightforward and stack one over the other. To enhance the shadow security, each shadow was encoded and decoded by HE method. With regards to the concern on image quality, the new Opposition-based Harmony Search (OHS) algorithm was used to produce the optimal key. The protected mechanism used was Searchable Symmetric Encryption (SSE) in view of blockchain. In addition, the client in this plan hadn't emphasized for the outcomes locally, in the study conducted by Li et al. in 2019 [18]. A chaos-based probabilistic block cipher for image encryption was proposed by Sakshi Dhall et al. in 2018 [19] in which the Random Bits insertion made the scheme probabilistic. This stage likewise helped in expanding the entropy and making the power circulation increasingly uniform in cipher. The created ciphertext was double the span of plain content. An expansion in ciphertext space was unavoidable for the probabilistic encryption since it leveraged the clear message space for the assailant got expanded. Three essential parts of the cloud are execution, accessibility, and security. Thus, there is the requirement for proficient homomorphic crypto algorithms. This research work by A. M. Vengadapurvaja et al. in 2017 [20] proposed a proficient homomorphic encryption algorithm to encrypt the therapeutic images in order to perform valuable activities on it without breaking the classification. In order to expand the security of encryption and decryption processes, the optimal key was chosen utilizing hybrid swarm optimization, i.e., grasshopper optimization and Particle Swarm Optimization in elliptic bend cryptography. In this perspective, the medicinal images were anchored in IoT structure (Mohamed Elhoseny et al. in 2018) [21]. In this execution, the outcomes were thoroughly analyzed, while a different encryption algorithm with its optimization techniques was also tested.

4.3 Purpose of Digital Image (DI) Security in WSN

- Nowadays DI security in WSN is vital for wide range of applications. So numerous researchers are investigating different encryption as well as decryption procedures [22, 23].
- Keys in encryption and decryption are computationally infeasible to infer the first image. Time stamps pinpoint the proprietor of the information and the time at which the information was produced.
- Recently, cryptographical technique is used in DI security process due to its image security. The frameworks used for encryption can be said as an assurance gadget for the secret image.
- The encryption is the place where the plain information can be changed over to figure or guaranteed information, and it can peruse just by decrypting it.
- A few security issues were identified with sophisticated image planning and transmission due to which it is vital to keep up the uprightness and the protection of the image.
- There is assorted securable image encryption that can be particularly for assurance against the unapproved get to and the security board is utilized to have the client's confirmation, exactness in information security of WSN.

4.4 Methodology

In the proposed model, the DI in WSN is encrypted by utilizing LWC-hash function with optimal key given for the cipher image. The primary thought is that an image can be seen as a course of action of blocks. The coherent data present in an image is because of the relationship among the image components in a given course of action. The proposed hash function makes use of the idea of changing over an unverifiable measure of digital information to a preset measure of information, through rolling out minor improvements in input information features, bringing about key changes in yield information. Hash functions ought to acknowledge the images of any length as input, deliver a yield of settled length at faster rates. In addition, the optimal key arrangement utilizes ECS optimization technique in the hash function. The security and proficiency of the image in WSN thoroughly depend on the inherent cipher which needs confused calculation.

Benefits of the proposed hash function in LWC

- One of the major utilizations of hashing is that one can view two records for equality. Without opening the two report records to think about it, the determined hash estimations of these documents enable the proprietor to know quickly in the event that they are extraordinary.
- It is ought to be equipped for restoring the hash of an input rapidly. On the off chance, the procedure isn't quick enough and the framework just won't be effective.
- Cryptographic hashes take clear content passwords and transform it into an enciphered image for capacity. Anyway, the fact is that the trust in existing encryption algorithms can be exchanged to hash value. It is difficult to express such a favorable position in monetary terms though it positively affects the alternative of a hash function.

4.4.1 Hash Functions for DI Security

A cryptographic hash function [24] licenses one to promptly affirm that some information coordinates store hash esteem, yet makes it difficult to reproduce the data from the hash alone. The proposed image security approach considers the hash function with three features given below which needs to be fulfilled for this examination.

Preimage Resistance (PR): This element has pre indicated yields with unimaginable input function with appropriate hash values. For example, impressive image I, hash function H and the hash estimation of image security are $H(I)$. A hash function is said to be pre-image safe when an assault is performed against it.

Second Preimage Resistance (SPR): It is computationally difficult to locate any second input which has indistinguishable yield from any predetermined information I. So the hash estimation of this SPR is $H_1(I) = H_2(I)$ denoting that the hash function authentication includes 2n work.

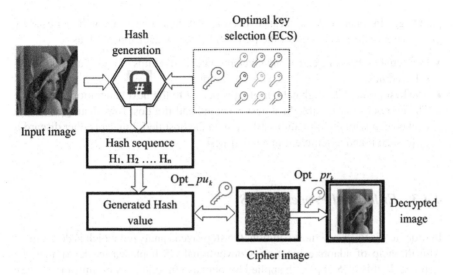

Fig. 4.1 Hash function structure

Collision Resistance (CR): A hash impact occurs when two arbitrary images, I_1 *and* I_2 hash to similar esteem. So the hash esteem is determined by $H(I_1) = H(I_2)$ $\because I_1 \neq I_2$. In addition, it is appropriate to alternate the relations in a hash function and furthermore execute multi-block collision to allude two impacting messages, each comprising of somewhere around two blocks.

A cryptographic hash function is a scientific algorithm that takes a subjective size of information and encodes it to a settled size of information, commonly close to 128 bits. The output digital images and the model of hash function are illustrated in Fig. 4.1. Rather than utilizing a hash function with variable size information, a function with settled size information is made and utilized a vital number of times.

The hashing sequence of the image can be attained in the acceptor side while the securing process can be accomplished by looking at the similarity of the received hashing sequence and the hashing succession produced from the received image. Robustness implies that the hashing esteems have not changed after traditional attacks, which guarantee the ordinary image security over in WSN.

4.4.2   Enhanced Cuckoo Search (ECS) Model

With increasing advancements nowadays, naturally-propelled metaheuristic algorithms [25] are broadly used to solve hard optimization issues. These algorithms depend on arbitrary Monte-Carlo method, guided by some nature-enlivened insight, especially development and swarm knowledge. In general, cuckoos detain its prepared eggs in other cuckoos' nests expecting that their off-springs will be raised by proxy parents [26]. For most of the part, CS has a few decisions such as

- Each cuckoo lays one egg at any given moment and dumps its egg in an arbitrarily-selected nest
- The best home with high caliber of eggs persist to the next generation
- The number of accessible host nest is settled and if a host fowl distinguishes the cuckoo egg with the probability [0, 1], then the host fledgling can either discard it or forsake it and assemble it in another nest.

4.4.3 Enhanced Process of CS

In order to discover new nest solution, the step size is analyzed which is determined with the help of adaptive function. Conventional CS haphazardly determines the step size. In this ECS approach applied for our hash function key optimization, every egg is positioned and characterized into the best gathering with better quality eggs, and a relinquished gathering with more awful quality eggs. The flow diagram of the proposed ECS is shown in Fig. 4.2.

ECS is effectively used to tackle planning issues and used to solve design optimization issues in auxiliary designing and in various other applications like speech reorganization, job scheduling, global optimization.

Step 1: Key initialization

The key matrix is initialized with N dimensions, so the size of the initialize solution is $K = \{K_1, K_2, \ldots K_N\}^T$ and $K_i = \{K_i^1, K_i^2, K_i^3 \ldots K_i^M\}$. Each solution randomly generates the objective which is calculated as Peak Signal to Noise Ratio (PSNR).

Step 2: New key solution by updating procedure

The updating technique is then performed by taking the regard fitness whereas the Levy flight condition is utilized to find the nest position. In this examination, the standard ECS algorithm enhances its execution. In this circumstance, making new answers K_{new} for a cuckoo Levy flight joining with the inertia weight, controls the chasing limit. This is performed and defined by

$$K_{n(new)}^i = K_n^i + \alpha * r * Levy(\beta) \tag{4.1}$$

$$\therefore \quad Levy(\beta) = t(-\beta), \quad 1 < \beta < 2 \tag{4.2}$$

In light of the above process, the new solutions are created and the step size is discovered with the enhanced procedure. Each egg in a nest indicates a solution and a cuckoo egg indicates another solution where the goal is to supplant the weaker fitness solution by another solution.

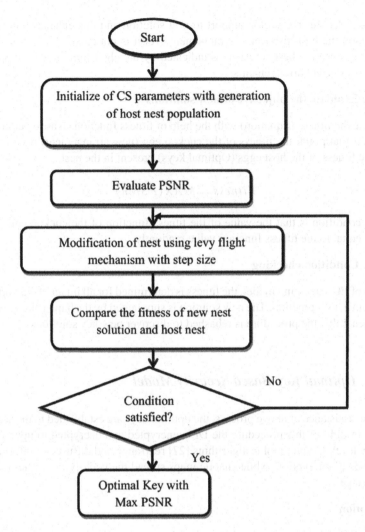

Fig. 4.2 Flow chart for ECS

Step 3: Calculate the step size

This step size ought to be identified with the sizes of the issue, the algorithm which is endeavoring to understand. The estimation of α affects the last solution since it prompts diverse new solutions and is set to various qualities. The equation for step measure is

$$Stepsize = 0.01 \frac{k^{t+1}}{|v^{t+1}|^{1/\beta}} (v - k_{best}) \qquad (4.3)$$

Here v denotes the vector support to nest solution and this enhanced algorithm relies upon the best-discovered arrangement. Yet it is not recalled in any different memories since the best solution, as indicated by the algorithm, is constantly held among the current arrangements.

Step 4: Evaluate the objective of new solutions

Compute the fitness (Equation) with the help of fitness function so as to get optimal value. At that point, the fitness of the cuckoo egg (new arrangement) is contrasted with the fitness of the host eggs (Optimal keys) present in the nest.

$$Fitness = MAX(PSNR) \tag{4.4}$$

The condition is that the value of the fitness function of the cuckoo egg is less than or equal to the fitness function value randomly.

Step 5: Condition checking

In light of ECS updating model, the fitness is determined for all images using optimal keys with a hash function. The new solution is supplanted by the haphazardly-picked nest. Generally, the procedure is rehashed to discover new key solutions.

4.4.4   Optimal Key-Based Security Model

With the assistance of above process, the optimal keys are established to the security model. In light of this procedure the DI is encrypted and decrypted to upgrade the security level. Cryptographic algorithms [27] rearrange and diffuse the information by rounds of encryption, while chaotic maps spread the underlying locale over the whole stage.

Encryption

Assume 256 pixel plain images in terms of the matrix with rows and columns while the range of pixel value is between 0 and 255 values. Based on this, the hash value is generated for the assumed input image. It is scientifically represented by

$$HASH(En(Image, opt\_pu_k), H_k) = H(image, H_k)$$
$$HASH(En(Image, opt\_pu_{k1}), H_k) = HASH(En(Image, opt\_pu_{k2}), H_k)$$
$$\tag{4.5}$$

In Eq. (4.5), pu_{k1}, pu_{k2} are optimal encryption keys and with the help of this, the image is divided into blocks; each of dimension 16×16. After dividing, there will be total 256 such blocks. The algorithm uses various averages when encrypting different input images to even with the same sequence based on hash function.

Decryption

The optimal private key is generated with the diffusion model. The security of the cipher should only rely on the decryption keys, pr_{k1}, pr_{k2} since an adversary can improve the plain image from the observed cipher image once gets pr_{k1}, pr_{k2}.

$$Decrypted\ Image = Decrypt(C) \qquad (4.6)$$

$$HASH(Dec) = C \oplus \mod(H_k + Cipher, 256) \oplus pr_{k1}, pr_{k2} \qquad (4.7)$$

The decryption process is administered by at least one cryptographic key. As a rule, the key utilized for the procedure of decryption and depiction is not really indistinguishable, depending on the framework utilized.

Overall steps of the proposed security model
Input: Digital Plain Image
Output: Secured Image
{
Let Assume DI (0 o 255 pixels) in the matrix
Find the Hash value of DI
Optimize pr_k, pu_k of security Model
{
Initialize Key matrix
New solution(Keys) generated by Levy flight Process
Calculate the Step size of the Levy flight
Evaluate the fitness of image
Its maximum the process will be an exit
Otherwise
Again iteration = iteration + 1
End
}
Optimal pu_k key based Encryption
Cipher to Plain image by optimal pr_k
}

4.5 Result and Analysis

This optimal key-based hash function cryptography was implemented in MATLAB 2016 with an i5 processor and 4 GB RAM. For this evaluation process, the measures considered were PSNR, MAE, Entropy, NPCR, and throughput. The proposed model was compared with other encryption techniques.

Tables 4.1 and 4.2 tabulates the performance and image results for the proposed model. Here, the table represents the encrypted image with histogram and

Table 4.1 Security measures for optimal hash function

Images	PSNR (dB)	Entropy	MAE	NPCR (%)	Throughput (Mb)
Lena	59.52	7.99	0.29	99.45	24.45
Barbara	49.45	7.94	0.27	96.22	29.45
Baboon	43.22	7.99	0.26	97.55	19.21
House	53.22	7.89	0.33	96.48	32.11
Airplane	56.122	7.99	0.29	98.12	27.48
Cameraman	60.47	7.95	0.16	98.49	26.22

the decrypted images dependent on the optimal key-based encryption framework. The five different images (Lena, Barbara, Baboon, house, airplane, and cameraman) were analyzed by the proposed encryption and decryption model. For instance, the cameraman image values of PSNR and MAE were 60.47 and 0.16 respectively. The image encryption is said to be imperative when the encrypted image ought not to be conspicuous.

The comparative analyses of security modeling with different measures is illustrated in Fig. 4.3. Here Fig. 4.3a demonstrates the PSNR rate of just hash function, hash-CS system and hash-OCS strategy with all images. The most extreme PSNR of the proposed model was 59.44 dB in baboon images when compared with other images. Then Fig. 4.3b demonstrates the entropy esteem, its dimension which is not exactly an encrypted image and therefore it can go under entropy attacks effectively, while the encrypted image can withstand the entropy assaults because of a more elevated amount of entropy values. In addition, the rest of the charts (c), (d), (e) discuss the MAE, NPCR and throughput values. When the extensive rate of NPCR was estimated, then it demonstrated that the situation of pixels got haphazardly changed. This specific change in estimation of pixels makes it exceptionally hard to distinguish between the unique and encrypted images.

Table 4.3 demonstrates the MIN, MED and MAX PSNR estimations of ordinary optimization and the proposed ECS procedures. The minimum values of CS and ECS were 2.04 and 2.28 respectively which met the greatest PSNR estimation of cipher secured images. In addition, the graphical representation of this fitness evaluation is shown in Fig. 4.4.

The proposed ECS algorithm is near to a hypothetical optimal value and the precise of the algorithm got enhanced. In the meantime, through the numerical result of standard deviation, the dependability of the ECS algorithm is found to be truly great. From this diagram, it can be inferred that the optimization impact of ECS algorithm was better (59.48 dB) than CS algorithm. These curves of CS algorithm start to straighten demonstrating the fact that CS algorithm has trapped local optimum PSNR value.

Time intricacy and throughput of the security model is shown in Table 4.4. Here various DIs are viewed, for example, Lena, Barbara, baboon, airplane, house and cameraman images. The encryption time is utilized to compute the throughput of

Table 4.2 Image-based results for ECS-Hash cryptography technique

Original image	Size	Cipher image	Histogram for cipher image	Decrypted image
	871*532			
	871*532			
	871*532			
	871*532			
	96x93			
	128x128			

Table 4.3 Fitness (PSNR) evaluation

Technique	Minimum PSNR	Median PSNR	Maximum PSNR
CS	2.04	21.24	52.222
ECS	2.28	26.85	59.48

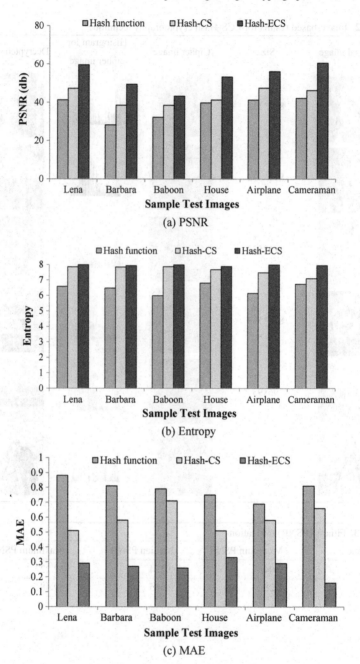

(a) PSNR

(b) Entropy

(c) MAE

Fig. 4.3 Security measure comparative analysis

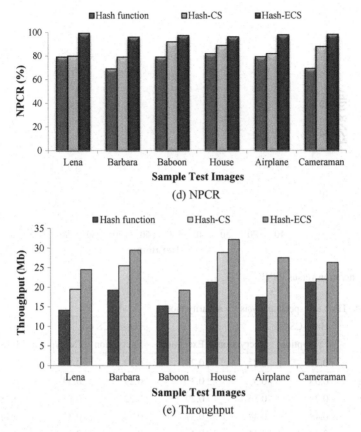

(d) NPCR

(e) Throughput

Fig. 4.3 (continued)

an encryption system. The throughput of the encryption method was determined by separating the size of the document by absolute encryption time in second. It is better to be accomplished in a hash with an optimal key by ECS. For instance, the house images took 0.33 and 0.44 s for encryption and decryption in ECS. These values were determined for cipher and plain images opposite to pixel network amid the security model.

Figure 4.5 demonstrates the security level examination of all the considered images with various methodologies. Among those strategies, the security level accomplished in hash function with ECS optimization approach is also considered. The security level was found to be 86.14–94.56% in all images around and for higher security, the optimal key was used.

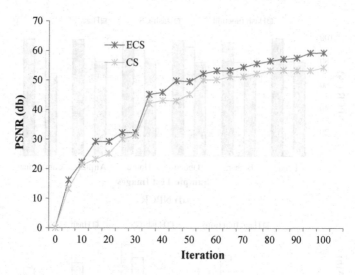

Fig. 4.4 Iteration versus PSNR

Table 4.4 Time and speed analysis for security model

Images	Hash-CS (s)			Hash-ECS (s)		
	Encryption	Decryption	Execution	Encryption	Decryption	Execution
Lena	0.21	0.22	0.28	0.19	0.19	0.22
Barbara	0.33	0.3	0.3	0.29	0.24	0.3
Baboon	0.29	0.36	0.29	0.22	0.17	0.28
House	0.48	0.44	0.37	0.38	0.33	0.44
Airplane	0.39	0.41	0.43	0.21	0.32	0.36
Cameraman	0.44	0.52	0.51	0.27	0.42	0.4

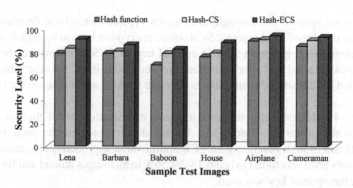

Fig. 4.5 Security level analysis

4.6 Conclusion

In this chapter, a novel image security-based optimal hash function was proposed in wireless networks. This model exchanges the images securely through WSN and makes it conceivable to encrypt the information in real-time applications because of its low computational expense. The proposed encryption strategy in this investigation was tested on various configuration images which demonstrated great outcomes. For performance examination, impressive measurements such as PSNR, MAE, Entropy, NPCR, and throughput were considered. In light of these parameters, the assessment of these input images, ciphered by an optimal public key and decrypted by an optimal private key, gave promising results (94.56%) in terms of secure dimension in hash function with ECS approach. The entropy of encrypted image in any of the other competitive algorithms can be consequently fused during the encryption process of any images in WSN. In the future, another optimization model can be used to secure the DI over sensor networks with the help of other symmetric and asymmetric encryption strategies.

References

1. Thakur, S., Singh, A. K., Ghrera, S. P., & Elhoseny, M. (2018). Multi-layer security of medical data through watermarking and chaotic encryption for tele-health applications. *Multimedia Tools and Applications*, pp. 1–14.
2. Avudaiappan, T., Balasubramanian, R., Pandiyan, S. S., Saravanan, M., Lakshmanaprabu, S. K., & Shankar, K. (2018). Medical image security using dual encryption with the oppositional based optimization algorithm. *Journal of Medical Systems, 42*(11), 208.
3. Shankar, K., & Eswaran, P. (2016). RGB-based secure share creation in visual cryptography using optimal elliptic curve cryptography technique. *Journal of Circuits, Systems, and Computers, 25*(11), 1650138.
4. Shankar, K., Elhoseny, M., Chelvi, E. D., Lakshmanaprabu, S. K., & Wu, W. (2018). An efficient optimal key based chaos function for medical image security. *IEEE Access, 6*, 77145–77154.
5. Bansod, G., Pisharoty, N., & Patil, A. (2016). PICO: An ultra lightweight and low power encryption design for ubiquitous computing. *Defense Science Journal, 66*(3).
6. Thorat, C. G., & Inamdar, V. S. (2018). Implementation of new hybrid lightweight block cipher. *Applied Computing and Informatics*.
7. Ratha, P., Swain, D., Paikaray, B., & Sahoo, S. (2015). An optimized encryption technique using an arbitrary matrix with probabilistic encryption. *Procedia Computer Science, 57*, 1235–1241.
8. Weng, L., & Preneel, B. (2011, October). A secure perceptual hash algorithm for image content authentication. In *IFIP International Conference on Communications and Multimedia Security* (pp. 108–121). Berlin, Heidelberg: Springer.
9. Agrawal, H., Kalot, D., Jain, A., & Kahtri, N. (2014, July). Image encryption using various transforms-a brief comparative analysis. In *2014 Annual International Conference on Emerging Research Areas: Magnetics, Machines and Drives (AICERA/iCMMD)* (pp. 1–4). IEEE.
10. Elsayed, W., Elhoseny, M., Sabbeh, S., & Riad, A. (2017). Self-maintenance model for Wireless Sensor Networks. *Computers & Electrical Engineering*.
11. Usha, M., & Prabhu, A. (2018, June). Performance Analysis of Encryption Algorithms with Pat-Fish for Cloud Storage Security. In *International Conference on Mobile and Wireless Technology* (pp. 111–120). Springer, Singapore.

12. Balouch, Z. A., Aslam, M. I., & Ahmed, I. (2017, April). Energy efficient image encryption algorithm. In *2017 International Conference on Innovations in Electrical Engineering and Computational Technologies (ICIEECT)* (pp. 1–6). IEEE.
13. Nguyen, D. Q., Weng, L., & Preneel, B. (2011, October). Radon transform-based secure image hashing. In *IFIP International Conference on Communications and Multimedia Security* (pp. 186–193). Berlin, Heidelberg: Springer.
14. Mohd, B. J., & Hayajneh, T. (2018). Lightweight block ciphers for IoT: Energy optimization and survivability techniques. *IEEE Access, 6,* 35966–35978.
15. Riad, A., Elhoseny, M., Elminir, H., & Yuan, X. (2016). A secure data routing schema for WSN using elliptic curve cryptography and homomorphic encryption. *Journal of King Saud University-Computer and Information Sciences, 28*(3), 262–275.
16. Shih, H., Ho, J., Liao, Y., & Pan, J. (2013). Fault node recovery algorithm for a wireless sensor network. *IEEE Sensors Journal, 13*(7), 2683–2689.
17. Shankar, K., Elhoseny, M., Kumar, R. S., Lakshmanaprabu, S. K., & Yuan, X. (2018). Secret image sharing scheme with encrypted shadow images using optimal homomorphic encryption technique. *Journal of Ambient Intelligence and Humanized Computing*, 1–13.
18. Li, H., Tian, H., Zhang, F., & He, J. (2019). Blockchain-based searchable symmetric encryption scheme. *Computers & Electrical Engineering, 73,* 32–45.
19. Dhall, S., Pal, S. K., & Sharma, K. (2018). A chaos-based probabilistic block cipher for image encryption. *Journal of King Saud University-Computer and Information Sciences.*
20. Vengadapurvaja, A. M., Nisha, G., Aarthy, R., & Sasikaladevi, N. (2017). An efficient homomorphic medical image encryption algorithm for cloud storage security. *Procedia Computer Science, 115,* 643–650.
21. Elhoseny, M., Shankar, K., Lakshmanaprabu, S. K., Maseleno, A., & Arunkumar, N. (2018). Hybrid optimization with cryptography encryption for medical image security in the Internet of Things. *Neural Computing and Applications*, 1–15.
22. Shankar, K., & Eswaran, P. (2015). A secure visual secret share (VSS) creation scheme in visual cryptography using elliptic curve cryptography with an optimization technique. *Australian Journal of Basic and Applied Sciences, 9*(36), 150–163.
23. Oad, A., Yadav, H., & Jain, A. (2014). A review: Image encryption techniques and its terminologies. *International Journal of Engineering and Advanced Technology (IJEAT)*, pp. 2249–8958.
24. Weng, L., & Preneel, B. (2011, October). A secure perceptual hash algorithm for image content authentication. In *IFIP International Conference on Communications and Multimedia Security* (pp. 108–121). Springer, Berlin, Heidelberg.
25. Zhang, M., He, D., & Zhu, C. (2016, December). A cuckoo search algorithm based on hybrid-mutation. In *2016 12th International Conference on Computational Intelligence and Security (CIS)* (pp. 538–542). IEEE.
26. Begum, A. A. S., & Nirmala, S. (2018). Secure visual cryptography for a medical image using a modified cuckoo search. *Multimedia Tools and Applications, 77*(20), 27041–27060.
27. Shankar, K., & Eswaran, P. (2016). A new k out of n secret image sharing scheme in visual cryptography. In *2016 10th International Conference on Intelligent Systems and Control (ISCO)*, IEEE (pp. 369–374).

Chapter 5
An Optimal Haar Wavelet with Light Weight Cryptography Based Secret Data Hiding on Digital Images in Wireless Sensor Networks

Abstract Security is the rising concern in this specialized rebellion that attracts the analysts towards research and new commitment in Wireless Sensor Network (WSN) field. This chapter proposes a creative technique for image security in WSN using Steganographic and cryptographic model which secures the selected cover images and secret information. The effective Opposition-based Particle Swarm Optimization (OPSO) with Haar wavelet coefficients from Discrete Wavelet Transform (DWT) was brought into the embedding procedure. From this procedure, the encrypted file was deciphered though the encoded document may hide the information even now. This optimal wavelet attained the most extreme Hiding Capacity (HC) and PSNR rate. Finally, the stego images were considered in the security Model i.e., LWC-based SIMON block cipher. It works on the basis of key and round generation model and towards the end, the reverse procedure occurs with the image decryption and extraction modeling. The usage results demonstrated that the proposed security strategy has the most extreme CC and PSNR values (52.544) with minimum error rate (0.493) in comparison with other conventional strategies.

Keywords Steganography · Security in WSN · Discrete wavelet transform (DWT) and particle swarm optimization

5.1 Introduction

Information security is the base for dealing secret information in WSNs. In a heterogeneous WSN, notwithstanding the system organizing factors, example distance to the base-station, also distance among nodes, factors, for example, introductory energy, information processing capacity [1], able to fill in as a group head, and node versatility incredibly impact the system life expectancy. The heterogeneous model is an adjusted model of homogeneous clustering model, i.e., LEACH [2]. Steganography and cryptography are two different ways of accomplishing the transmission of secret information. Steganography is not the same as cryptography [3]. The primary motivation behind steganography is to pass on the information secretly by hiding

© Springer Nature Switzerland AG 2019   65
K. Shankar and M. Elhoseny, *Secure Image Transmission in Wireless Sensor Network*
(WSN) Applications, Lecture Notes in Electrical Engineering 564,
https://doi.org/10.1007/978-3-030-20816-5_5

specific attendance of information in some other medium, for example, image [4]. In WSN security model, steganography is not implied as a substitute for cryptography though the extended information can be encrypted and afterwards secretly imparted by means of Steganography which implies to include privacy in WSN [5]. The ongoing patterns and advancements in information technology emphasize the requirement for sheltered, secured and ensured transmission of data. The traditional encryption techniques lack in the ideal outcome i.e., securing the data [6]. Numerous applications utilize wavelet decomposition in general. Some parts of these applications are compression and denoising i.e., some of the spatial area procedures are LSB, PVD, EBE, RPE, PMM and Pixel intensity-based and so on whereas frequency space system portions are DCT, DWT, DFT, IWT, and DCVT [7].

The critical parts of the spatial area image exist in the approximation band that comprises of low frequency, edge and texture details and for the most part it exist in high-frequency subgroups [8]. 2D DWT could be accomplished specifically through lifting-based plan and fast convolution-based plan [9]. Regular lifting-based models require less or few arithmetic activities when compared with convolution-based methodology for DWT and they have long basic ways [10]. The safe embedding and extracting procedures ought to include encrypting and decrypting a mystic message with an amazingly-solid cryptographic algorithm and secure appropriation of cryptographic algorithm key [11]. Asymmetric 64-bit block cipher was created by Bruce Schneier which was further improved for 32-bit processors with substantial data reserves. It is essentially quicker than DES, AES, etc. [12]. PSO is a populace-based algorithm on the basis of birds flock. The abuse and investigation qualities are adjusted by inactivity weight [13]. This security model is utilized for the exchange of images over web where the image security turns into real security for military, security offices, social or portable applications [14].

5.2   Literature Review

Author/Year/Ref. number	Techniques	Description	Measures
Elhoseny et al. [15]	Genetic algorithm based method that optimizes heterogeneous sensor node clustering in WSNs	The proposed method greatly extends the network life, and the average improvement with respect to the second best performance based on the first-node-die	NLT, PDR

(continued)

(continued)

Author/Year/Ref. number	Techniques	Description	Measures
Elhoseny an Hassanien [16]	Genetic algorithm (GA) based clustering over WSNs	This work is to form the network structure that optimize its throughput with maximum NLT	Network performance, NLT and throughput
Nipanikar et al. [17]	DWT with PSO	This research focused on image steganography utilizing sparse description and an algorithm named PSO algorithm for viable determination of the pixels to embed the secret sound signal in the image. It is based on pixel choice method utilizing a fitness function that relies upon cost function	PSNR, MSE, entropy
Sidhik et al. [18]	Haar wavelet	The transform domain technique (wavelet transform) was utilized to accomplish high capacity alongside security. Further, the key element of this work is that it retained the nature of the cover image which goes about	MSE, PSNR, WPSNR (weighted), structural content (SC)
Thanki and Borra [19]	DWT and finite ridgelet transform (FRT)	FRT was connected on the cover color image to get ridgelet coefficients of each color channel of a cover color image and a solitary dimension DWT was connected to get distinctive wavelet coefficients	Normalized correlation (NC), PSNR, MSE, hiding capacity
Shankar et al. [20]	DWT with oppositional based harmony search	The encrypted shadow can be recuperated just by picking some subsets of these 'n' shadows that makes it straightforward and stack over one another. To enhance the shadow security, each shadow was encrypted and decoded utilizing HE method	PSNR, entropy, MAE, MSE

(continued)

(continued)

Author/Year/Ref. number	Techniques	Description	Measures
Nipanikar and Deepthi [21]	Haar wavelet	A strategy for hiding the text message in the image was proposed for which a DWT was utilized with the cost function that finds a situation to experience embedding	PSNR, MSE, correlation
Valandar et al. [22]	Integer wavelet transform (IWT) and modified logistic chaotic map	A new transform domain steganography strategy, dependent on IWT, was investigated for digital images and it utilized a riotous map. This map is a changed logistic map which builds the key length and security of the proposed strategy	PSNR, MSE, structural similarity (SSIM) index
Yuvaraja and Sabeenian [23]	DWT with fuzzy	Fuzzy rules were developed and utilized to distinguish the edges in both cover and mystery images. The development of fuzzy rules streamlined the edge location process by finding the thin and thick edges in both cover and secret images	PSNR, MSE, MAE, entropy, Bhattacharyya coefficient and NC

5.3 Purpose of Steganography in WSNs

- In general, the motivation behind steganography in WSN is to keep up the secret communication between two gatherings [24–26].
- The reason to stego an image is that one need anon-excess portrayal of the image. Moreover, the first and stego images must be of similar size.
- In wavelet transform, the sizes in the term of width and the function of the wavelet get changed with each spectral segment.
- Embedding is completed by altering the least noteworthy bits of chosen wavelet coefficients. Accordingly, the number of fluctuating coefficients for this situation is equivalent to the number of capable coefficients.
- Hiding information, particularly in steganography systems, has been contemplated inside and out than the non-vigorous strategies.

5.4 Steganography with Cryptography Model

The Digital Image security model-based secret information hiding in WSN is imperative to process in the current situation. In this model, it develops the limit, power and security. So it is a need to utilize the wavelet transform-based steganography and LWC strategy. In the implementation of proposed security work over WSN, the first step is that the secret information needs to be embedded with cover images by making use of ideal DWT model. Image transform coefficients are coefficients that have expansive scale and perceptually critical in image representation. Leaving aside the ideal wavelet coefficients, the research still utilizes the swarm-based OPSO system. Once the embedding process is completed, the secret key is used to encrypt and decrypt the images with most extreme security level in WSN. At last, the IDWT is applied to decrypt the images so as to extricate the cover image and secret information. The detailed summary is discussed in the sections below. The image transmission over WSNs is done by the routing protocol AODV.

5.4.1 Embedding Model: Wavelet Transform

DWT transforms the discrete domain to frequency image of Stego Images (SI). It clearly divides high from the low-frequency information based on pixels. When embedding the cover image with secret information, the SI gets decayed by four groups, in view of the lower and higher bands i.e., LL, HH, LH, and HL. It has some noteworthy features. This transformation model, Haar wavelet, is used to distinguish the coefficient that utilizes OPSO optimization methods. With the assistance of optimal coefficients, LL band images make SI protected from different assaults though it can prompt twists in stego image. The DWT is mathematically spoken as

$$I(t) = \sum_{m \in a} \sum_{n \in a} (I, \alpha_{m,n}) \cdot \alpha_{m,n}(t) \tag{5.1}$$

For most of the part, wavelet decomposition depends on filter banks. In general, the wavelet decomposition and reconstruction structures comprise of filtering. In this model, the principle goal of steganography is to keep some unintended spectator away from stealing or decimating the secret information.

(1) **Haar wavelet model**

Haar wavelet is used to discover the coefficients of image decomposition process. It relies on the image pixel grid (0–255), and for the most part it has two arrangements of coefficient vectors i.e., H1 and H2. Discrete Haar functions may be portrayed as functions constrained by looking at the Haar limits. This function of Haar mother wavelet appears as α

$$\alpha = \begin{cases} 1 & 0 \le t \le 1/2 \\ -1 & 1/2 \le t \le 1 \\ 0 & otherwise \end{cases} \tag{5.2}$$

The functions possess the property such that each function is persistent on interim [0, 1] and might be spoken by a consistent and convergent series as far as the components of this framework, Wavelets, are concerned. For example, those Haar wavelet transform conserve the strength of transformation process. In order to improve the compression ratio of the proposed model, the Haar wavelet coefficient is optimized using OPSO technique. In order to enhance the PSNR of LL band images, the OPSO model is used to optimize the coefficients.

5.4.2   OPSO Model for DWT

To enhance the PSNR and throughput values, optimal wavelet selection model was considered in this research work. The level decomposition is essentially associated with the LL band of the present decay arrange, which structures a recursive decomposition system as shown in Fig. 5.1. OPSO method was used to enhance the wavelet coefficients with the resistance procedure and is represented by

$$O_i = B_j + A_j - e_j \tag{5.3}$$

The fitness function examines each produced solution selected arbitrarily. The development starts with complete irregular arrangement of elements and is rehashed in consequent generations. The objective function is represented by

$$PSNR = 10 \cdot Log_{10}\left(\frac{Max_I^2}{MSE}\right) \tag{5.4}$$

The objective function condition (5.4) is assessed by the Image grid M * N for the intensity values to embedding model. The created particles are generally shown

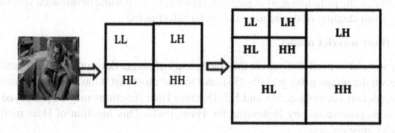

Fig. 5.1 DWT decomposition process

to the data. From this PSNR rate, the cover image with secret image was implanted utilizing the ideal coefficient vectors.

Initialization of Wavelet coefficients

Here initialize the particles solution and opposition solutions (coefficients vectors) arbitrarily on the DWT matrix which can be determined by the search position i.e.,

$$\beta_{i,j} = I(v_{i,j})/NF \quad \because \quad NF = \sum_M \sum_N I(v_{i,j}) \tag{5.5}$$

Initialization equation (5.5) $I(v_{i,j})$ is a variation of intensity values whereas ZF is considered as normalized vector values.

Construction and updating model

On account of this initialization process, the probability values are determined in order to estimate the intensity of considered images with OPSO updating technique. The reasons for updating model is to locate the ideal coefficient vectors with the most extreme PSNR value. Here two essential factors are announced i.e., *P_best and G_best* values. The velocity and position of the particles in the first OPSO are given as:

$$PV_i(t+1) = PV_i(t) + d_1 rand\,(P\_best(t) - r_i(t)) + d_2 rand(G\_best - r_i(t)) \tag{5.6}$$

$$PV_{i+1} = r_i(t) + PV_i(t+1) \tag{5.7}$$

In Eqs. (5.6) and (5.7), $PV_i \rightarrow$ particle velocity, $r_i \rightarrow$ the current position of a particle, rand is a random number between (0, 1) and $d_1, d_2 \rightarrow$ learning factor, usually $d_1 = d_2 = 2$. This updating method works when updating the new coefficient vectors of Haar wavelet to build the hiding limit of the algorithm when compared with different frameworks. However, by this technique, the computational complexity is minimally high.

End Criteria: When the most extreme PSNR, with optimal coefficients, is achieved in DWT, the process is finished. Otherwise, the updating procedure of OPSO is rehashed. In view of the optimal Haar wavelet, the image and secret information are used for embedding and extrication processes.

5.4.3 *Embedding Process:* **Generation of Stego Image (SI)**

- Separate the image into "N" number of blocks and take the secret information. Play the haar wavelet decomposition.
- Apply transforms' domain strategy upon a cover grayscale image and secret grayscale image. By applying DWT, the approximation coefficients of the matrix such as LH, HL, HH, and LL are extricated.

- In order to diminish the extra information, all the more explicitly, quantization table is to be shared between the sender and the recipient.
- To conceive the coefficients in order, the pseudorandom permutation scheme is connected with ideal coefficient vectors and the stego image is created using IDWT.

5.4.4   Extraction Algorithm

- Apply IDWT upon secured stego image in order to remove bit sequence.
- Optimal coefficients are decided on the basis of extracting secret message bits from the integer coefficients. The removed bits are additionally changed over into its unique secret information.
- Separate the wavelet coefficients and obtain DWT for the fused image to recreate the secret image. Figure 5.2 shows the embedding and extraction Stego Images.
- The private key is helpful to decode the cover image and secret information of the embedded image in the wake of applying DWT.

5.4.5   Security Model: LWC

The LWC attempts to address this by proposing algorithms and protocols that are planned explicitly to perform well on these obliged stages. In this chapter with regards to the upgradation of SI security, the SIMON block ciphers were considered. The selected block cipher was executed well alike block cipher-based hash work since it is powerful to be associated in hardware. And further it is sufficient enough to adapt and perform well on full range of the obliged stages which spurred the researchers to pick the least difficult part conceivable. Every family includes ten unique block

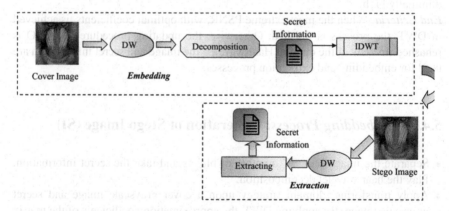

Fig. 5.2 Stego image embedding and extracting

ciphers with varying block and key sizes to firmly fit the application necessities. A noteworthy key size and block size of the SIMON cipher is from 32 to 128 and 64 to 256 respectively.

Features of SIMON block cipher

a. SIMON block cipher is an intrigue of S-boxes and it is utilized as a piece of a Substitution-Permutation Network (SPN) to keep the assault.
b. To build the security on one-time work which is important to perform the crypt-analysis so as to reduce the time taken for encryption and decoding works.
c. Lightweight block ciphers frequently use bit permutations as a feature of SPN. The function of these bit permutations is to spread bits around in some optimal selection.
d. An individual key differential trademark and an individual key differential is present in 15-round SIMON48, a lightweight block cipher.
e. The key of a block cipher uses reliable properties of key schedules which are created upon the number of pixels in images.

5.4.6 Block Cipher Designing and Security Model

Stego image security model introduces the keys with few conditions coupled with round-based encryption and encryption model. For most of the part, the 2n-bit blocks of the image are considered and are represented by

$$E(SI) = Cipher^i_{q_l}, \ldots Cipher^n_{q_1}, \quad i > 1 \tag{5.8}$$

The usage function of the ciphers $C^i_{q_l}$ is called round functions. Likewise, the separate keys are round keys. When the functions are identical (or practically indistinguishable), such a cipher is likewise alluded as an iterated block cipher.

Key generation: The key structure could be adjusted and conceived. The keyword count is utilized to decide the structure of the key development bringing about a total bit width of the SI.

The key sizes provide with progressively-noticeable security, yet decrease the encoding and decoding rates. The 64 bit-key sizes is rarely used whereas the regularly utilized is 128-bit key size. Further, the block size and key size are shown in Table 5.1.

The specification supports the utilization of these constants as a means of dispensing with sliding properties and circular shift symmetries between the distinctive round keys. Bitwise XOR is implied as $a \oplus b$ whereas the correct bitwise revolution ROR is implied as $s^{-c}(a)$ in which c is the pivot count. The key generation is as follows

$$Key_{i+m} = \begin{cases} Ci \oplus k_i \oplus (SI \oplus S^{-1})(S^{-3}k_{i+1}) \ r = 2 \\ Ci \oplus k_i \oplus (SI \oplus S^{-1})(S^{-3}k_{i+2}) \ r = 3 \\ . \\ . \\ . \\ Ci \oplus k_i \oplus (SI \oplus S^{-1})(S^{-3}k_{i+n}) \ r = n \end{cases} \tag{5.9}$$

The motivation behind the key sequence is to kill slide properties. Further, a bit mask of an XOR "1" exists against the most minimal two bits of the watchword which is incentive in the conditions.

Round function: The structure of the round function is indistinguishable for all renditions of the cipher. The round function works on two blocks of 'n' bits in order to make a single-encoded SI. The contributions of 128-bit plaintext and 128-bit key to create 128-bit cipher message occur in 68 rounds. This procedure is depicted through the following steps

- Bitwise AND movement is performed on arbitrary two bits of n-bit words.
- XOR is performed on the eventual outcome of bitwise AND task. One is the bit from lower block and the last value is XOR with one of the random pieces from upper block ultimately in XOR with a key.
- Left bitwise upset rounds are implied as $R^y(SI)$ in which 'y' is the rotation count.

For encryption, the SIMON round capacity is detailed as follows

$$RF(a_l, a_r, k_{round}) = ((R^1(a_l) \ \& \ R^8(a_l)) \oplus R^2(a_l) \oplus a_r \oplus k_{round}, w_l) \tag{5.10}$$

For Decryption

$$RF^{-1}(a_l, a_r, k) = (a_r, (R^1(A_l)\&r^8(a_l)) \oplus r^2(a_r) \oplus a_l \oplus k_{round}) \tag{5.11}$$

It creates a bit field that makes use of the following block by a left roll of 1 and 8, with a logical AND, operating on the present block as an XOR. Encryption is embed-

Table 5.1 Specification for SIMON cipher	Block size	Key size	Number of rounds
	32	64	32
	48	72, 96	36, 36
	64	96, 128	42, 44
	96	96, 144	52, 54
	128	128, 192, 256	68, 69, 72

ded in applications or accommodation functions that prohibit the implementation of hardware. In light of this procedure, the stego image is encrypted and decrypted and finally during extraction process, the IDWT is connected to decrypted images.

5.5 Result and Analysis

The proposed steganography with LWC model was executed in MATLAB 2016 with a system configuration of i5 processor with 4 GB RAM. This enhanced security process contrasted with other wavelet transform and encryption models with various measures, for example, PSNR (db), MSE and Hiding Capacity (byte).

Table 5.2 presents the list of images such as the cover images, secret information image process and the optimal haar wavelet-based stego images. Further, Fig. 5.3 demonstrates the computational time of embedding, extraction and security processes. The secret information and color images took practically 86.12% of the entire computational time. This progression time should be decreased in future with the help of advanced encryption algorithms and machines with high processing speed coupled with high physical memory. The most extreme times taken for embedding and extraction model of house image were 26.22 and 23.14 respectively and there is an effect on speed of the procedure time for the proposed algorithm.

Tables 5.3 and 5.4 demonstrate the results of the proposed procedure (steganography and cryptography) images and performance measures. For instance; in Lena image, the PSNR was 56.22, MSE (0.52), CC (0.99) and the hiding capacity of secret information was 79.65%. Likewise different parameters in all the images improved the consequences of steganography with LWC model.

Table 5.2 Steganography images (Lena, Barbara, Baboon and House)

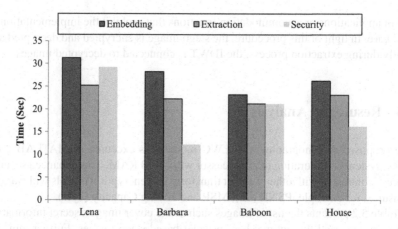

Fig. 5.3 Time analysis

Table 5.3 Image results for optimal DWT with Simon cipher

| L L s t e g o I m a g e | C i p h e r I m a g e | H i s t o g r a m | Decrypted Image |

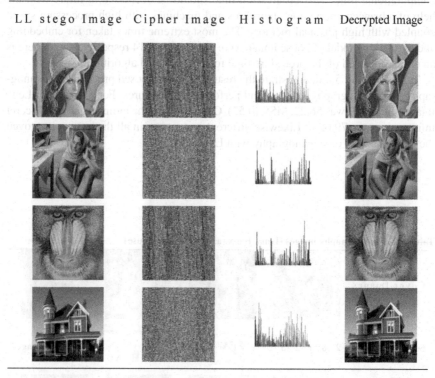

Table 5.4 Security measure results for steganography with cryptography

Images	PSNR (db)	MSE	CC	Hiding capacity (byte)
Lena	56.22	0.52	0.99	79.65
Barbara	53.222	0.99	1	83.56
Baboon	49.45	0.188	0.98	81.56
House	51.285	0.274	1	87.77

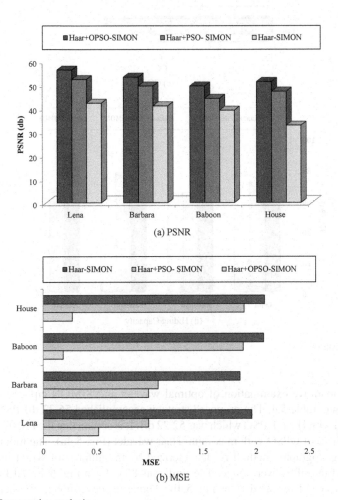

(a) PSNR

(b) MSE

Fig. 5.4 Comparative analysis

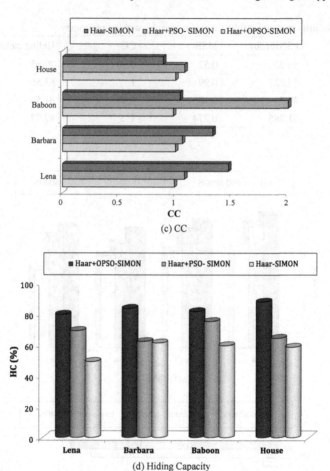

(c) CC

(d) Hiding Capacity

Fig. 5.4 (continued)

A comparative examination of optimal wavelet and SIMON ciphers appear in Fig. 5.4 and Table 5.4. The proposed strategy accomplished 56.22 dB PSNR when compared with Haar + PSO which was 52.22 dB. It is certain that the PSNR achieved with the help of OPSO with maximum Haar wavelet (Fig. 5.4a). The hiding capacity of steganography method is characterized by the number of secret image bits embedded into the cover image. The limit was 83.65% in Fig. 5.4d. At that point, the MSE rate got appeared in the Fig. 5.4b whereas the cover image, the stego image and the esteem 255 remain the most extreme pixel size of the image. At long last, the

Correlation Coefficient (CC) is shown in the Fig. 5.4c. The CC values for the image security in case of Baboon images for various methods were Haar + OPSO-SIMON (0.188), Haar + PSO-SIMON (1.88) and Haar-SIMON 2.07.

5.6 Conclusion

This chapter detailed about an innovative Steganographic method with LWC model for image security in WSNs. The selected optimal DWT coefficients were utilized that transformed the image from spatial domain to frequency domain through binary transformation whereas the binary-converted text was implanted in the image over WSN. Haar with OPSO model is accustomed to find directions of pixels which are utilized in embedding and extracting forms. The experimental results of stego image generation of HC were compared with conventional haar wavelet model. In addition, it provided high security for steganography and cryptographic models. Cipher image was created for LL band images whereas the SIMON block cipher was utilized. The primary favorable position of this cipher was the block size and key generation is one of its kind. The test results demonstrated that high PSNR, CC and Hiding limit values i.e., 52.54 db, 0.99 and 83.15 respectively with least MSE i.e., 0.493. The future augmentation is suggested to perform advanced optimization and Discrete Cosine Transform (DCT) for stego image process and symmetric encryption systems in WSN.

References

1. Elhoseny, M., Elleithy, K., Elminir, H., Yuan, X., & Riad, A. (2015). Dynamic clustering of heterogeneous wireless sensor networks using a genetic algorithm, towards balancing energy exhaustion. *International Journal of Scientific & Engineering Research, 6*(8), 1243–1252.
2. Elhoseny, M., Hassanien, A. E. (2019). Secure data transmission in WSN: An overview. In *Dynamic wireless sensor networks. Studies in systems, decision and control* (Vol. 165, pp. 115–143). Springer, Cham.
3. Hemalatha, S., Acharya, U. D., & Renuka, A. (2015). Wavelet transform based steganography technique to hide audio signals in image. *Procedia Computer Science, 47,* 272–281.
4. Muhammad, N., Bibi, N., Mahmood, Z., Akram, T., & Naqvi, S. R. (2017). Reversible integer wavelet transform for blind image hiding method. *PLoS ONE, 12*(5), e0176979.
5. Thanikaiselvan, V., Arulmozhivarman, P., Subashanthini, S., & Amirtharajan, R. (2013). A graph theory practice on transformed image: A random image steganography. *The Scientific World Journal, 2013.*
6. Shet, K. S., & Aswath, A. R. (2015). Image steganography using integer wavelet transform based on color space approach. In *Proceedings of the 3rd International Conference on Frontiers of Intelligent Computing: Theory and Applications (FICTA) 2014* (pp. 839–848). Springer, Cham.

7. Priya, A. (2018). High capacity and optimized image steganography technique based on ant colony optimization algorithm. *International Journal of Emerging Technology and Innovative Engineering, 4*(6).

8. Sharma, V. K., Mathur, P., & Srivastava, D. K. (2019). Highly secure DWT steganography scheme for encrypted data hiding. In *Information and Communication Technology for Intelligent Systems* (pp. 665–673). Springer, Singapore.

9. Subramanian, M., & Korah, R. (2018). A framework of secured embedding scheme using vector discrete wavelet transformation and lagrange interpolation. *Journal of Computer Networks and Communications, 2018*.

10. Shankar, K., & Lakshmanaprabu, S. K. (2018). Optimal key based homomorphic encryption for color image security aid of ant lion optimization algorithm. *International Journal of Engineering & Technology, 7*(1.9), 22–27.

11. Shankar, K., & Eswaran, P. (2016). RGB-based secure share creation in visual cryptography using optimal elliptic curve cryptography technique. *Journal of Circuits, Systems and Computers, 25*(11), 1650138.

12. Avudaiappan, T., Balasubramanian, R., Pandiyan, S. S., Saravanan, M., Lakshmanaprabu, S. K., & Shankar, K. (2018). Medical image security using dual encryption with oppositional based optimization algorithm. *Journal of Medical Systems, 42*(11), 208.

13. Elhoseny, M., Shankar, K., Lakshmanaprabu, S. K., Maseleno, A., & Arunkumar, N. (2018). Hybrid optimization with cryptography encryption for medical image security in Internet of Things. *Neural Computing and Applications*, 1–15.

14. Sathesh Kumar, K., Shankar, K., Ilayaraja, M., Rajesh, M. (2017). Sensitive data security in cloud computing aid of different encryption techniques. *Journal of Advanced Research in Dynamical and Control Systems, 9*, 2888–2899.

15. Elhoseny, M., Yuan, X., Yu, Z., Mao, C., El-Minir, H., & Riad, A. (2015). Balancing energy consumption in heterogeneous wireless sensor networks using genetic algorithm. *IEEE Communications Letters, IEEE, 19*(12), 2194–2197.

16. Elhoseny, M., Hassanien, A. E. (2019). Optimizing cluster head selection in WSN to prolong its existence. In *Dynamic wireless sensor networks. studies in systems, decision and control* (Vol. 165, pp. 93–111). Springer, Cham.

17. Nipanikar, S. I., Deepthi, V. H., & Kulkarni, N. (2017). A sparse representation based image steganography using particle swarm optimization and wavelet transform. *Alexandria Engineering Journal*.

18. Sidhik, S., Sudheer, S. K., & Pillai, V. M. (2015). Performance and analysis of high capacity steganography of color images involving wavelet transform. *Optik-International Journal for Light and Electron Optics, 126*(23), 3755–3760.

19. Thanki, R., & Borra, S. (2018). A color image steganography in hybrid FRT–DWT domain. *Journal of information security and applications, 40*, 92–102.

20. Shankar, K., Elhoseny, M., Kumar, R. S., Lakshmanaprabu, S. K., & Yuan, X. (2018). Secret image sharing scheme with encrypted shadow images using optimal homomorphic encryption technique. *Journal of Ambient Intelligence and Humanized Computing*, 1–13.

21. Nipanikar, S. I., Deepthi, V. H. (2017). Entropy based cost function for wavelet based medical image steganography. In *2017 International Conference on Intelligent Sustainable Systems (ICISS)* (pp. 211–217). IEEE.

22. Valandar, M. Y., Ayubi, P., & Barani, M. J. (2017). A new transform domain steganography based on modified logistic chaotic map for color images. *Journal of Information Security and Applications, 34*, 142–151.

23. Yuvaraja, T., & Sabeenian, R. S. (2018). Performance analysis of medical image security using steganography based on fuzzy logic. *Cluster Computing*, 1–7.

24. Shankar, K., Elhoseny, M., Chelvi, E. D., Lakshmanaprabu, S. K., & Wu, W. (2018). An efficient optimal key based chaos function for medical image security. *IEEE Access, 6*, 77145–77154.

25. Shankar, K., Lakshmanaprabu, S. K., Gupta, D., Khanna, A., & de Albuquerque, V. H. C. (2018). Adaptive optimal multi key based encryption for digital image security. *Concurrency and Computation: Practice and Experience*, e5122. https://doi.org/10.1002/cpe.5122.
26. Shankar, K., & Eswaran, P. (2017). RGB based multiple share creation in visual cryptography with aid of elliptic curve cryptography. *China Communications, 14*(2), 118–130.

Chapter 6
An Optimal Singular Value Decomposition with LWC-RECTANGLE Block Cipher Based Digital Image Watermarking in Wireless Sensor Networks

Abstract Numerous watermarking applications require implanting strategies that supply power against normal watermarking attacks, similar to pressure, noise, sifting, and so on. Dense sending of wireless sensor networks in an unattended situation makes sensor hubs defenseless against potential assaults. With these requests, the confidentiality, integrity and confirmation of the imparted data turn out to be important. This chapter investigated the optimal Singular Value Decomposition (SVD) strategy which was proposed by utilizing the Opposition Grey Wolf Optimization (OGWO) system for image security in WSN. This is a protected method for watermarking through the installed parameters required for the extraction of watermark. The objective function is utilized, at the optimization procedure, through which the greatest attainable robustness and entropy can be attained without debasing the watermarking quality. When the optimal parameters such as 'K', 'L' and 'M' got the images installed with secret data, at one point, the Light Weight Cryptography (LWC)-RECTANGLE block cipher process was used to encrypt and decrypt the watermarked images, transmitted in WSN. This encryption procedure has two critical procedures such as key generation and round function. The adequacy of the proposed strategy was exhibited by comparing the results with traditional procedures with regards to the watermarking performance.

Keywords Watermarking · Optimization · LWC · Block cipher · Security · Robustness · WSN

6.1 Introduction

With the development of World Wide Web, one can possibly do anything but it is difficult to disperse and transmit the multimedia information like images, sound and video from one place to the next with only a single snap [1]. Advanced image watermarking process gives copyright insurance for the image information by covering up fitting data in the first image [2]. We at that state express the basics of information encryption in remote sensor systems, demonstrating promising methodologies when managing image sensing [3]. As per the requirement for unique information amid

© Springer Nature Switzerland AG 2019   83
K. Shankar and M. Elhoseny, *Secure Image Transmission in Wireless Sensor Network (WSN) Applications*, Lecture Notes in Electrical Engineering 564,
https://doi.org/10.1007/978-3-030-20816-5_6

the watermark recognition processes, computerized watermarking can be character-ized as private as well as public calculations [4]. In view of the strategies utilized for watermark installation and [5] extraction, undetectable watermarking procedures are of three types such as spatial domain, frequency domain, and mixed domain. Imper-ceptible watermarking is an optimization issue [6]. Leaving it aside, the watermark shows up when the hues are isolated for printing. Spatial area processing includes the expansion of fixed amplitude pseudo-noise into the image [7]. The watermark can be covered up, into the information, so as to expect that the Least Significant Bit (LSB) information is outwardly unnecessary [8]. The routing procedures and remote sensor arrange demonstrating are getting much inclination, the security issues are yet to get broad core interest in WSN [9].

The watermark is inserted in DFT, DCT and DWT coefficients and this type of model is robust against basic image handling activities like low pass separating, contrast and brightness modification, and so on. In any case, they are hard to execute as well as computationally and progressively exorbitant [10]. During the installation, the watermark might be encoded into cover information by making use of a particular key [11]. This key is utilized to encrypt the watermark as an extra assurance level that utilizes LWC in WSN [12]. In cryptography, the message is generally mixed up. In any case, when the correspondence occurs, it is noted down. In spite of the fact that the data is covered up in the cipher [13–16], an interference of the message can be damaging though it demonstrates everything that there is a correspondence between the sender and beneficiary. The measure of the block is 64 bits whereas the length of the key differs up to bits [17]. Authentication is additionally required for some applications, so as to guarantee that recovered data originates from legitimate source nodes [18, 19]. Further, the watermark embedders do not have a key to decrypt the plain content values to insert the watermark. Subsequently, watermarking in the encrypted domain is demanding [20].

6.2   Literature Review

Makbol et al. [21] recommended false positive issue and conducted a study in which they contemplated, examined and presented in detail. The distinctive plans are con-centrated and characterized on the basis of likelihood of presentation to the false positive issue. A wide range of SVD-based inserting calculations that prompt false positive issue and its related potential assaults were assessed with the help of depend-ability tests and all the answers for false positive issue were inspected.

Najafi and Loukhaoukha [22] investigated the SVD properties which is connected on both watermark and unique images. The SVD-based watermarking plans are not safe against equivocalness assaults whereas when one experiences the ill effects of the false positive issue, this complaint is resolved without adding additional means to the watermarking calculation and the recommended plan is safe and secure against uncertainty assaults. The simulation of the scheme was executed and its strength against different sorts of attacks was tested.

The ideal partial forces of the change and the inserting quality factor were assessed through a metaheuristic optimization algorithm to upgrade the watermark subtlety and robustness by Abdelhakim et al. [23]. The methodology enabled a safe method to watermark by inserting the parameters required for the watermark extraction. A fitness function was utilized at the optimization procedure through which the greatest reachable strength was provided without corrupting the watermarking quality underneath the predicted quality limit.

A digital watermark can be inserted in host information at its spatial area alike in recurrence space. In this research work, a hybridized method consolidating Discrete Wavelet Transform (DWT) and Singular Value Decomposition (SVD) was investigated by Poonam et al. [24]. The increased utilization of PCs, web and advanced multimedia innovation prompts an effective sharing of computerized information/media. Despite this, the accessibility of various image processing tools encourages unapproved utilization of such information.

Increasing the number of vehicles on roads prompts blockage and security issues. WSN is a promising innovation giving Intelligent Transportation Systems (ITS) to address these issues by Tarek Gaber et al. in 2018 [25]. A trust demonstrate is structured and used to process a trust level for every hub and the Bat Optimization Algorithm (BOA) is utilized to choose the group heads dependent on three parameters: lingering vitality, trust esteem and the quantity of neighbors. The reenactment results demonstrated that our proposed model is energy proficient.

As wavelet-based image watermarking works according to the human visual framework, it is gaining significant attention in ensuring the copyright data, opined by Deepa B. Maheshwari in 2018 [26]. A new watermark was created which was then implanted into the host image. The technique was examined against performing distinctive assaults, for example, revolution, expansion of Gaussian and Poisson noise, normal separation and so forth in the watermarked host image.

A watermarking method for a color image was presented on the basis of Discrete Wavelet Transform, Discrete Cosine Transform and Singular Value Decomposition (DWT-DCT-SVD) by Yuqi He and Yan Hu in 2018 [27]. DWT was connected to an illuminant part Y whereas the low recurrence was isolated into blocks by utilizing discrete cosine transform which led SVD with each block. At long last, the watermark was implanted on the cover image. The test results demonstrated that the calculation has great intangibility and strong robust, and can viably oppose normal watermark assaults.

6.3 Purpose and Benefits of Image Watermarking in WSN

- The watermarked image ought not to influence the nature of the first image; however along these lines it ought to be imperceptible for the human eye.
- Watermarking performed as per this method is robust against image processing tasks like low pass filtering, splendor, complexity modification, obscuring, and so on [7, 10].

- Watermarks cannot be viewed by ordinary eyes. Invisible watermarks are more secure, robust than visible watermarks and increase the security level of the secret information in networking model [13].
- Watermarks are not perceptible through a standard eye. Imperceptible watermarks are more secure, vigorous than unmistakable watermarks and can expand the security dimension of the secret data in WSN [13].
- It can be smashed without stretching by any signal processing assaults. Though it is not helpless against assaults and noise, it is especially impalpable.
- Embedding, performed at the perceptually-important part of the image, has its own focal points in light of the fact that most compression schemes evacuate the perceptually-irrelevant segment of the image [17].

6.4   Proposed Model

The objective of this chapter is to investigate the image security and hiding using watermarking as well as Lightweight Cryptography (LWC) technique in WSN system. This model embeds the secret information in the host image using spatial to frequency transform. Here, the optimization-based SVD model is used and the reason for embed and extraction is to increase the security as well as the robustness of the work.

6.4.1   Image Watermarking

In case of digital watermarking model, the secret information was inserted into a host image to create watermarked images. For this procedure, there is no requirement for additional storage space. This watermarked image was installed straightforward on the components of singular values in the first image's singular values as shown in Fig. 6.1.

The security level of the watermarking framework was improved whereas the OGWO optimization was used to locate the singular values. The watermarked bits were inserted on the intensity of each embedding and extricating blocks with low recurrence singular values. When the optimal parameters were recognized, the best arrangement were inserted and separated on the basis of secret messages. Besides this, the proposed model provided astounding visual nature of the watermarked images and attacked the connected process.

6.4.2 Optimization-Based Singular Value Decomposition (SVD)

SVD is an asymmetrical transform of numerical examination. Here, the study considered the lattice values which were decayed by Eigenvector as well as Eigenvalues, I. In general, the transform matrix gets decayed by three factors such as 'K', 'L' and 'M'. In light of these values, the image was inserted and then extracted for the security procedure. It is one of its attractive properties and the features of SVD incorporate its stability with little aggravation. By and large, it has some set of features such as,

- When an image is somewhat bothered, its singular values do not change fundamentally.
- The first-singular value in the value arrangement acquired by SVD [28] activity of an image is a lot bigger than the best ones.
- The reconstructed image quality is generally not debased if small singular values of its item are eliminated.
- The removed watermark image is most probably influenced by the procedure of geometry activities, particularly the extraction of the watermarking process.

In light of the SVD qualities, the host image can withstand certain geometric contortions when singular value decomposition is executed on the procedures of insertion

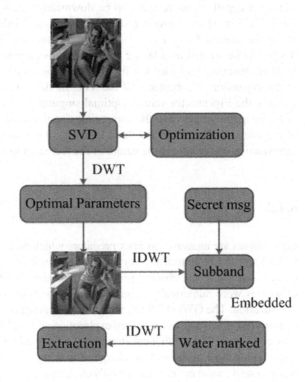

Fig. 6.1 Overview of the security model

as well as extraction of watermark. The decomposition examination considered the image 'P' with matrix R * C which can be scientifically communicated by

$$P = (KLM)^T \Rightarrow \sum_{i=1}^{n} K_i * L_i * M_i \qquad (6.1)$$

Here K and M are the orthogonal matrices of size R * C whose column vectors are left singular and right singular vectors, respectively. The terms K and M are unitary matrices whereas L is a diagonal matrix element. In general, the element P symbolizes the involvement of every layer of the decomposed image within the final image formation. The expansion of the SVD decomposed matrix is shown in Eq. (6.2).

$$P = KLM^T = \begin{pmatrix} K_{11} \dots K_{1R} \\ K_{21} \dots K_{R2} \\ \dots \\ \cdot \\ \cdot \\ K_{R1} \dots K_{RR} \end{pmatrix} * \begin{pmatrix} L_{11} \dots 0 \\ L_{21} \dots 0 \\ \dots \\ \cdot \\ 0 \dots L_{RC} \end{pmatrix} * \begin{pmatrix} M_{11} \dots M_{1R} \\ M_{21} \dots M_{R2} \\ \dots \\ \cdot \\ M_{R1} \dots M_{RR} \end{pmatrix} \qquad (6.2)$$

A matrix with positive diagonal elements was assumed from first to the last row in descending order. The diagonal components of L were named as LMs of P, which were non-negative and ventured to be downwardly-composed. So the matrix was $L1, L2, L_C$ in which the term L denotes the singular values of framework 'P', fulfilling the request $L_1 > L_2 > \cdots \geq L_C$. In watermarking grounded on SVD, an image can be treated as a lattice and can be disintegrated into three lattices. The SVD computation included the disclosure of the eigen values alike eigenvectors. In the expansion of robustness of the SVD model, OGWO technique was used to enhance the Eigenvector values. Optimal singular values, related to the luminance of the image brightness and the comparing singular vectors, indicated the geometry of the image. In the event that these little singular values were ignored during the reproduction of the image, the nature of the recreated image got corrupt [28].

6.4.3   OGWO

Grey wolves are measured as apex predators which means that they are at the peak in leading the life naturally. Grey wolves, for most of the part in their lives, wish to live as a pack. The chasing procedures and the social chain of the importance of wolves were numerically demonstrated so as to create GWO and perform the optimization. The GWO [29] calculation was attempted with standard test works to demonstrate that it possess unrivaled exploration and exploitation qualities than other swarm insights. The leading constraints are a male and a female, called alpha. The alpha is, for most of the part, in charge of making choices about chasing, sleeping place, time to wake up, etc. The alpha's choices are directed to the pack.

(i) ***Opposition process***

The generation of opposition-solution was demonstrated for singular value optimization on the basis of converse course of action. From which, the best solutions were picked up by taking a gander at the generated opposition solution [29]. This proposed methodology starts with a superior solution set and continues checking the contrary arrangement all the while in the inquiry space. It was performed until there is no enhancement in the best arrangement is possible. The flowchart of OGWO is shown in Fig. 6.2.

(ii) ***Objective function***

It is characterized as a proportion of the normal data substance in an Image. It is higher than that of the other existing techniques. Thus, this strategy provided the development of watermarked information with more data which is described in the condition (6.3).

$$En = \sum P(w_i) * \log_2 P(w_i) \tag{6.3}$$

First, in these findings, the probability values were determined for the watermarked images. So in order to reduce the complexity of the algorithm, the OGWO was consolidated with the utmost entropy in foreign fiber image watermarking process.

(iii) ***New singular values*** $(K, L$ *and* $M)$ ***updating process***

Three unique solutions were considered to the refreshing model such as α, β *and* δ which are described as first greatest, second best and third best arrangements. The underlying three best singular values achieved in the recent times require other interest authorities to change their circumstances, according to the circumstance of the best inquiry administrator. For repetition, the new solution $c(t + 1)$ was assessed using the formulae given below (6.4).

$$W^\alpha = |Q_1.a_\alpha - a|, \quad W^\beta = |Q_2.a_\beta - a|, \quad W^\delta = |Q_3.a_\delta - a| \tag{6.4}$$

$$W = |Q.a_P(t) - a(t)| \tag{6.5}$$

In the conditions (6.4) and (6.5), the new grey wolfs were updated for SVD transform process. With decreasing A, half of the accentuations got centered on the examination $(|Q| < 1)$ and the other half focused on the usage. Encasing the direct, the resulting conditions were utilized recalling a definitive target to give numerical model.

6.4.4 Security of Watermarked Images

The RECTANGLE design made use of the bit-slice strategy in a lightweight way so as to accomplish an exceptionally-minimal effort in equipment as well as extremely-aggressive execution in programming. For security, the 64-bit watermarked images

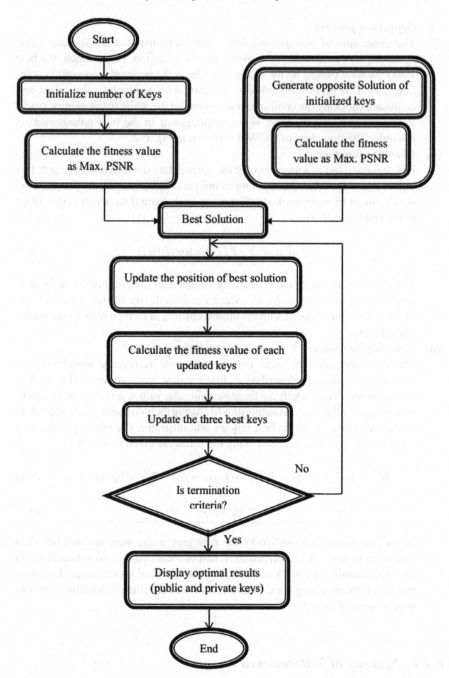

Fig. 6.2 Flowchart for OGWO

were considered from the middle of the result i.e., 64-bit cipher image. It indicated by $W = |w_1||w_2| \ldots |w_{63}|$ a cipher state with 16 bits. A 64-bit subkey was considered as a 4×16 rectangular cluster whereas the cipher image matrix is depicted through the condition (6.6). This block cipher additionally has two procedures such as key calendar and round schedule.

$$\begin{bmatrix} w_{0,15} \ldots w_{0,2} \; w_{0,1} \; \; w_{0,0} \\ w_{1,31} \ldots w_{1,18} \; w_{1,17} \; w_{1,16} \\ w_{2,47} \ldots w_{2,34} \; w_{2,33} \; w_{2,32} \\ w_{3,63} \ldots w_{3,50} \; w_{3,49} \; w_{0,48} \end{bmatrix} \quad (6.6)$$

(a) **Key schedule**

The keys were listed by 80 or 128 bits [30] and for 80-bit key, the seed key i.e., client-provided keys, were stored in the array bits. From these assortments of bits, a perfect key was selected to encrypt with the help of public key and decrypt using perfect private key of the watermarked image. The key matrix is shown in the condition below.

$$\begin{bmatrix} k_{15} \ldots k_2 \; k_1 \; \; k_0 \\ k_{31} \ldots k_{18} \; k_{17} \; k_{16} \\ k_{47} \ldots k_{34} \; k_{33} \; k_{32} \\ k_{63} \ldots k_{50} \; k_{49} \; k_{48} \end{bmatrix} \quad (6.7)$$

At last k25, the updated key state got extricated. At each round, five bits are moved to left side more than 1 bit with the new value being registered with new key generational values.

(b) **Round schedule**

RECTANGLE is a 25-round SP-organize cipher. Every round in it undergo three procedures such as (i) Add Round key, (ii) Sub column and (iii) Shift Row.

- A basic bitwise XOR of the round subkey to the middle state.
- Secondly, parallel use of S-boxes to 4 bits in a similar segment. The S-box-utilized RECTANGLE is a 4-bit to 4-bit S-box
- In the final process, a left revolution to every segment over different adjusts occurs when the Column 0 is not turned, then the row 1 is rotated to left side more than 1 bit whereas the row 2 is turned left side in excess of 12 bits while the line 3 is left turned in excess of 13 bits which are shown as the condition below (6.8).

$$\left(r_{0,15} \ldots r_{0,3} r_{0,2} r_{0,1} r_{0,0} \right) \xrightarrow{<<<0} \left(r_{0,15} \ldots r_{0,3} r_{0,2} r_{0,1} r_{0,0} \right)$$

$$(r_{1,15} \cdots r_{1,3}r_{1,2}r_{1,1}r_{1,0}) \xrightarrow{<<<1} (r_{1,14} \cdots r_{1,2}r_{1,1}r_{1,0}r_{1,15})$$
$$(r_{2,15} \cdots r_{2,3}r_{2,2}r_{2,1}r_{2,0}) \xrightarrow{<<<11} (r_{2,4} \cdots r_{2,8}r_{2,7}r_{2,6}r_{2,5}) \qquad (6.8)$$

The roundsubkey is made up of 4 initial lines of the present substance of the key [30]. Though it is very encouraging with increasing quantity of rounds, the minimum number of dynamic S-boxes achieved the greatest security for the watermarked images.

6.4.5   Watermarking Embedding Module

(i)   In the light of getting optimal values in SVD, the image and secret data were inserted whereas the extraction model invert process was directed.

(ii) Let us assume the host image is separated in 'N' blocks whereas all the sub-images have the same size with DWT connected process.

(iii) Consider each sub-image to have partitioned again in 8 * 8 blocks. At that point, the SVD is to be applied on the chosen block images.

(iv) For each chosen most noteworthy intricacy blocks in the main segment of P, the magnitude difference between the neighboring coefficients was determined. At long last, all the subband images were consolidated to apply IDWT technique. Finally, the watermarked image as retrieved.

6.4.6   Watermarking Extracting Module

(i)   The extractor evaluated the sum of the intensity values for blocking the host and watermarked image.

(ii) Perform Singular Value Decomposition to lower-high and high-lower bands.

(iii) Now remove half of the watermark image from each subband.

(iv) Apply DWT on all subband blocks relying on the sort of transform which had been connected before insertion.

(v)   Reconstruct the watermark image utilizing the removed watermark bits and the secured secret data while register the similitude between the first and extricated watermarks.

6.5   Results and Analysis

The proposed watermarking and the security model was implemented in MATLAB 2016a system configuration, i5 processors with 4 GB RAM. The performance of the security model was compared with other security models. The parameters considered were PSNR, Entropy, MSE, and NC.

Tables 6.1 and 6.2 show the results of various host images as well as secret information and the extensive hosts are (Lena, Barbara, Baboon, House and Airplane) for the proposed security model. Table 6.2 demonstrates that the watermarking is connected to install the watermark and remove the images. The security was evaluated based on time span taken by break the watermarking technique and uncover the hidden watermark. One favorable position of SVD-based watermarking is that there is no compelling reason to insert all the optimal singular values in a visual watermark. Contingent upon the extents of the biggest singular values, it seems to be adequate to install just a little set. The PSNR values of SVD-OGWO were 56.22, 53.22, 49.22, 50.14 and 47.11 dB individually. Another vital parameter of this watermarking security was Normalization Coefficient (NC) i.e., 1 whereas the current research chapter achieved 0.97 and 098, etc. Further, the MSE and the current study objective entropy a value distortion due to embedding the watermarkers kept at low dimension so that the PSNR of the watermarked image is moderate.

Figure 6.3 demonstrates the comparative analyses of all the measures in image security approach for the extensive systems such as SVD + GWO-RECTANGLE cipher, SVD + OGWO-RECTANGLE cipher and SVD-RECTANGLE cipher with all the parameters. The proposed mechanism, proved by the comparative analysis, is unraveled without doing additional works like authentication mechanism in singular-based arrangements. Figure 6.3a demonstrates the PSNR values which when compared with all methods, it showed 8–10% distinction. At that point, Fig. 6.3c demonstrates the NC values, watermarked essential watermark and watermarked host images which were found to be 0.99, 0.98, 0.96, 1 and 1 respectively. This indicates greater transparency of installed watermarks. The proposed methodology can enhance the heartiness degree by utilizing higher embedding strength that diminishes the watermarking quality. Table 6.3 demonstrates the followed and connected consequences of watermarking images i.e., salt and pepper noise is added to the watermark image. When applying the salt and pepper noise, the noise assaulted image gets separated from the watermark image. For instance, in baboon image, the PSNR of salt & pepper was 39.22 dB and it was 43.85 dB in pixel trade when compared with different image results. The watermarked image, proposed in this chapter, against various attack models, works commendably than the expected.

Table 6.1 PSNR, Entropy, HC, and CC of watermarked images

Images	PSNR	Entropy	MSE	NC
Lena	56.22	7.88	0.72	1
Barbara	53.22	7.95	0.26	0.96
Baboon	49.22	7.99	0.34	0.98
House	50.14	7.66	0.39	1
Airplane	47.11	7.99	0.41	1

Table 6.2 Image results

host image	Secret data	Embedded image (watermarked)	Extracted watermarked image

Table 6.3 Attack-based results analysis

Images	Attack	PSNR	Entropy	MSE	NC
Lena	Salt & pepper	52.66	6.89	0.99	1
Barbara		46.28	7.09	1.08	0.92
Baboon		39.22	7.85	0.85	0.96
House		32.24	7.09	2.05	1.09
Airplane		41.59	7.41	1.03	0.85
Lena	Pixel exchange	51.78	6.2	0.99	0.76
Barbara		39.22	7.11	1.05	1.08
Baboon		43.48	7.69	1.22	0.56
House		33.48	7.55	2.14	0.52
Airplane		32.22	7.28	2.03	0.44

6.6 Conclusion

This chapter proposed the hypothetical investigation of watermarking security model with SVD-OGWO decomposition module. The SVD-based watermarking strategy has indicated the power image quality against all the tests. In robustness analyses, we utilized OGWO process on SVD. According to the SVD-values, the compression attack got decreased by utilizing DWT and the robustness was added to the watermarked image. In order to get the watermarked image, LWC-RECTANGLE

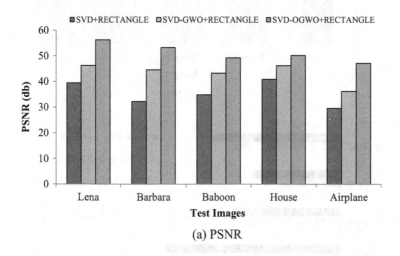

(a) PSNR

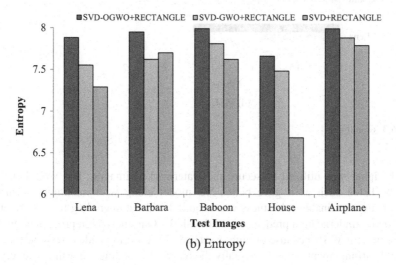

(b) Entropy

Fig. 6.3 Comparative analysis

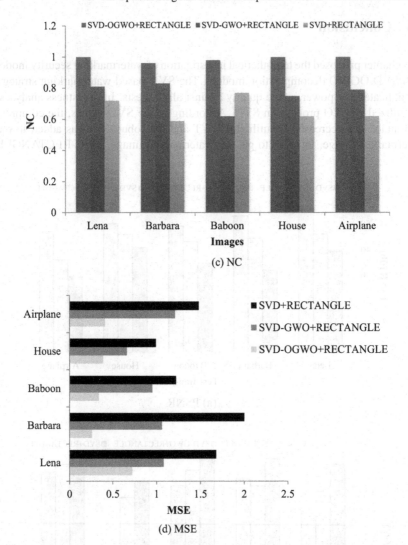

Fig. 6.3 (continued)

block cipher was utilized to secure the watermarked images. The SVD benefits can be listed such as it gave better NC model along with maximum security. The greatest attainable robustness was acquired without degrading the watermarking quality underneath a predetermined threshold. Our encryption process might be unfeasible for WSN because of high overhead of registering and correspondence; a possible arrangement could be actually the blend of encoding algorithm with cryptography. To assess the execution of the proposed technique, distinctive host images were chosen upon which a binary watermark was installed. From the implementation results, the PSNR, MSE, Entropy, and NC values were 51.182, 0.424, 7.894 and 0.988 respectively. In the future, other decomposition methods with innovative security schemes should be investigated. This outcome demonstrated that the proposed watermarking technique yielded high quality as well as robustness.

References

1. Fazli, A. R., Asli, M. E., Ghasedi, K., & aliZanjankhah, H. S. (2011, November). Watermarking security analysis based on information theory. In *2011 IEEE International Conference on Signal and Image Processing Applications (ICSIPA)* (pp. 599–603). IEEE.
2. Naveed, A., Saleem, Y., Ahmed, N., & Rafiq, A. (2015). Performance evaluation and watermark security assessment of digital watermarking techniques. *Science International, 27*(2).
3. El-Shorbagy, M. A., Elhoseny, M., Hassanien, A. E., & Ahmed, S. H. (2018). A novel PSO algorithm for dynamic wireless sensor network multiobjective optimization problem. *Transactions on Emerging Telecommunications Technologies*, e3523.
4. Song, C., Sudirman, S., Merabti, M., & Llewellyn-Jones, D. (2010, January). Analysis of digital image watermark attacks. In *Consumer Communications and Networking Conference (CCNC), 2010 7th IEEE* (pp. 1–5). IEEE.
5. Mahule, R. V., & Dhawale, C. A. (2015). Analysis of image security techniques using digital image watermarking in spatial domain. In *Nckite* (pp. 19–26).
6. Benrhouma, O., Hermassi, H., & Belghith, S. (2017). Security analysis and improvement of an active watermarking system for image tampering detection using a self-recovery scheme. *Multimedia Tools and Applications, 76*(20), 21133–21156.
7. Harjito, B., & Suryani, E. (2017, February). Robust image watermarking using DWT and SVD for copyright protection. In *AIP Conference Proceedings* (Vol. 1813, No. 1, p. 040003). AIP Publishing.
8. Shankar, K., Elhoseny, M., Chelvi, E. D., Lakshmanaprabu, S. K., & Wu, W. (2018). An efficient optimal key based chaos function for medical image security. *IEEE Access, 6,* 77145–77154.
9. Walaa, E., Elhoseny, M., Sabbeh, S., Riad, A. Self-maintenance model for wireless sensor networks. *Computers and Electrical Engineering* (In Press). Available Online December 2017.
10. Zhou, X., Zhang, H., & Wang, C. (2018). A robust image watermarking technique based on DWT, APDCBT, and SVD. *Symmetry, 10*(3), 77.
11. Eisenbarth, T., Kumar, S., Paar, C., Poschmann, A., & Uhsadel, L. (2007). A survey of lightweight-cryptography implementations. *IEEE Design & Test of Computers, 24*(6), 522–533.
12. Gupta, D., Khanna, A., Shankar, K., Furtado, V., & Rodrigues, J. J. Efficient artificial fish swarm based clustering approach on mobility aware energy-efficient for MANET. *Transactions on Emerging Telecommunications Technologies*, e3524. https://doi.org/10.1002/ett.3524.
13. Verma, V., Srivastava, V. K., & Thakkar, F. (2016, March). DWT-SVD based digital image watermarking using swarm intelligence. In *International Conference on Electrical, Electronics, and Optimization Techniques (ICEEOT)* (pp. 3198–3203). IEEE.
14. Shankar, K., & Eswaran, P. (2015). A secure visual secret share (VSS) creation scheme in visual cryptography using elliptic curve cryptography with optimization technique. *Australian Journal of Basic and Applied Sciences, 9*(36), 150–163.
15. Elhoseny, M., Shankar, K., Lakshmanaprabu, S. K., Maseleno, A., & Arunkumar, N. (2018). Hybrid optimization with cryptography encryption for medical image security in Internet of Things. *Neural Computing and Applications*, 1–15. https://doi.org/10.1007/s00521-018-3801-x.
16. Shankar, K., Elhoseny, M., Kumar, R. S., Lakshmanaprabu, S. K., & Yuan, X. (2018). Secret image sharing scheme with encrypted shadow images using optimal homomorphic encryption

technique. *Journal of Ambient Intelligence and Humanized Computing*, 1–13. https://doi.org/10.1007/s12652-018-1161-0.

17. Rajesh, M., Kumar, K. S., Shankar, K., & Ilayaraja, M. Sensitive data security in cloud computing aid of different encryption techniques. *Journal of Advanced Research in Dynamical and Control Systems, 18*.

18. Shankar, K., & Eswaran, P. (2016). An efficient image encryption technique based on optimized key generation in ECC using genetic algorithm. *Advances in Intelligent Systems and Computing, 394,* 705–714.

19. Elhoseny, M., Yuan, X., Yu, Z., Mao, C., El-Minir, H., & Riad, A. (2015). Balancing energy consumption in heterogeneous wireless sensor networks using genetic algorithm. *IEEE Communications Letters, IEEE, 19*(12), 2194–2197.

20. Shankar, K., & Eswaran, P. (2015). Sharing a secret image with encapsulated shares in visual cryptography. *Procedia Computer Science, 70,* 462–468.

21. Makbol, N. M., Khoo, B. E., & Rassem, T. H. (2018). Security analyses of false positive problem for the SVD-based hybrid digital image watermarking techniques in the wavelet transform domain. *Multimedia Tools and Applications*, pp. 1–35.

22. Najafi, E., & Loukhaoukha, K. (2019). Hybrid secure and robust image watermarking scheme based on SVD and sharp frequency localized contourlet transform. *Journal of Information Security and Applications, 44,* 144–156.

23. Abdelhakim, A. M., Saad, M. H., Sayed, M., & Saleh, H. I. (2018). Optimized SVD-based robust watermarking in the fractional Fourier domain. *Multimedia Tools and Applications,* pp. 1–23.

24. Poonam, & Arora, S. M. (2018). A DWT-SVD based robust digital watermarking for digital images. *Procedia Computer Science, 132,* 1441–1448.

25. Gaber, T., Abdelwahab, S., Elhoseny, M., Hassanien, A. E. Trust-based secure clustering in WSN-based intelligent transportation systems. *Computer Networks*. Available online September 17, 2018.

26. Maheshwari, D. B. (2018, February). An analysis of wavelet based dual digital image watermarking using SVD. In *2018 International Conference on Advances in Communication and Computing Technology (ICACCT)* (pp. 69–73). IEEE.

27. He, Y., & Hu, Y. (2018, May). A proposed digital image watermarking based on DWT-DCT-SVD. In *2018 2nd IEEE Advanced Information Management, Communicates, Electronic and Automation Control Conference (IMCEC)* (pp. 1214–1218). IEEE.

28. Pomponiu, V., & Cavagnino, D. (2011). Security analysis of SVD-based watermarking techniques. *International Journal of Multimedia Intelligence and Security, 2*(2), 120–145.

29. Mirjalili, S., Mirjalili, S. M., & Lewis, A. (2014). Grey wolf optimizer. *Advances in Engineering Software, 69,* 46–61.

30. Zhang, W., Bao, Z., Lin, D., Rijmen, V., Yang, B., & Verbauwhede, I. (2015). RECTANGLE: A bit-slice lightweight block cipher suitable for multiple platforms. *Science China Information Sciences, 58*(12), 1–15.

Chapter 7
Optimal Data Hiding Key in Encrypted Images Based Digital Image Security in Wireless Sensor Networks

Abstract An effective optimal Data Hiding (DH) key with Light Weight Encryption (LWE) is proposed in this chapter. There are numerous issues in wireless sensor organize like client authentication just as information travel in the system isn't so much secure. The image was encrypted with a secret key; yet conceivable to insert extra information for secret key. On sender side, the cover image was considered for the encryption procedure by hash function after which the data hiding method was used to hide the images utilizing the optimal data hiding keys. For key selection the current research utilized the Enhanced Cuckoo Search Optimization (ECS) method. The most extreme embedding capacity and security level were achieved. In reversible process, at the receiver side, the embedded image undergoes decryption procedure to recouped image and the data is selected by public key (Shankar and Eswaran in J Circuits Syst Comput 25(11):1650138, 2016) [1]. It can be applied in various application situations; moreover, the data extraction and image recovery are free of any errors. From this security model, the hided image containing the installed information can be then sent to combination focus utilizing sensor networks. At the combination focus, the hidden information is separated from the picture, the required handling is performed and choice is taken consequently. From the implementation procedure, the proposed model attained the most extreme Hiding Capacity (HC) (91.073%) and the greatest PSNR (59.188 dB), compared to other conventional techniques.

Keywords Data hiding (DH) · Key selection · Optimization · Security · Embedding · Extracting and light weight encryption

7.1 Introduction

Digital images have expanded quickly due to which the security of multimedia information is essential for some applications such as classified transmission, video observation, military and medicinal field applications in WSN [2]. Data Hiding (DH) or information hiding can embed extra information into cover information including content, sound, etc. [3]. In general, the information hider is not equivalent to infor-

© Springer Nature Switzerland AG 2019   99
K. Shankar and M. Elhoseny, *Secure Image Transmission in Wireless Sensor Network (WSN) Applications*, Lecture Notes in Electrical Engineering 564,
https://doi.org/10.1007/978-3-030-20816-5_7

mation proprietor and creating DH plans for encrypted images is considered essential [4]. Irreversible DH plans cannot be utilized to secure the information in encrypted images since irreversible [3]. DH may bring changeless contortion to the encrypted image which may prompt mistakes during the decryption stage. In this way, the reversible DH play a significant role in hiding data among encrypted images in encrypted space techniques [5]. WSN that assemble visual information from an observed field work under similar standards [6, 7], yet visual information detecting, preparing and transmission are all the more difficult because of the colossal measure of data to be taken care of when contrasted with scalar information [8].

During the time of installation, the pixel values relating to the peak point are either kept unaltered or expanded by one as indicated by secret bits. It is clear that the complete hiding payload of secret bits rely upon the pixel number of the peak point in the histogram [9]. One can easily extract the secret bits called as 'receiver' through the histogram of stegoimage and further move back the moved receptacles for image recuperation [10]. This DH procedure includes steganography, watermarking and some other techniques [11, 12]. With significant watermarking, it is inserted in an essential image into an extent such that it is discernible to a human observer. Though [13] the embedded information is not distinguishable, if there should be an occurrence of undetectable watermarking, it may very well be extricated by a computer program [13]. WSN in fundamental applications, one should think about the confinements of the identical. One of the prime concerns is data security [14]. The information that the sensor gets, ought to be taken care of and transmitted to the sink, which is then given to the base station and is open for the end customers by methods for the WSN [15].

A cryptographic hash function's current usage shows that it no more does the trick performed by advanced security objectives [16]. Two recent lightweight cryptographic natives are PRESENT block cipher and PHOTON hash function [17]. The test with the structure of lightweight hash functions is managing the harmony between security prerequisite and memory necessities [18]. Most designers of lightweight hash functions focus on the security prerequisite by delivering a yield size of no less than 256 bits to prevent any collision. Based on the hash function decoding stage, the DH key serves to extricate the extract information and the first image was recreated with gratitude to a median filter on the installed image [19]. This chapter discusses about the image security with DH installation and extraction process and the enhancement of security LWE strategy i.e., a hash function is used.

7.2   Literature Review

Concealing data in an image, in such a way that, it does not influence the first cover image pixels or cause a perpetual contortion after extracting that data is known as reversible data hiding process which was innovated by Zhang in 2011 [20]. This method hides two watermarks in a given encrypted image. The primary watermark is embedded by replacing the chosen encrypted image pixels dependent on DH key. The data hider does not realize the first image content which may reversibly implant the secret information into image and the difference is dependent on two-dimensional contrast histogram changes. Information extraction is totally distinct from image decryption i.e., information extraction is possible either in the encrypted space or decrypted area with the goal that it may very well be connected to various application situations, as proposed by Dawen Xu et al. in 2018 [21]. The strategy can insert secret data into 2 × 2 image blocks by misusing the pixel repetition within each block. By stretching out this idea to the encrypted area, a reversible DH technique was proposed in encrypted images utilizing adaptive block-level forecast error extension (ABPEE-RDHEI) [22]. A novel encryption method to verify information transmission in WSN with dynamic sensor clusters [23]. Every sensor and joins with hub ID, separation to the bunch head, and the list of transmission round to frame novel 176-bit encryption keys. Utilizing select OR, substitution, and stage activities, encryption and decryption are accomplished efficiently. Due to adaptive pixel determination and iterative inserting forms, the proposed ABPEE-RDHEI can accomplish a high installing rate and satisfying visual nature of the stamped decrypted images. In 2018, Qin et al. [24] recommended the information hider which classifies encrypted blocks into two sets compared to smooth and complex districts in a unique image. With DH key, the save space is emptied to suit extra bits by compacting LSBs of the block set related to the smooth district. The divisible tasks of information extraction, coordinate decryption, and image recuperation are directed by the receiver as indicated by the accessibility of encryption key as well as DH key. Lightweight cryptography is developed to upgrade the security level in inescapable processing applications, for example, those portrayed by smart, yet asset-compelled gadgets by Hammad et al. [25]. The K-coverage problem is a period and energy expending process and the association between sensors is required constantly. To address this issue, this article proposes a K-inclusion demonstrate dependent on genetic algorithm to expand a WSN life-time [26]. In the scan for the ideal dynamic spread, diverse factors, for example, targets' positions, the normal devoured vitality, and inclusion scope of every sensor are considered in WSN system. Some lightweight cryptographic hash functions are depicted from both equipment and programming points of view. In the examination of these techniques, the distinction of these native and customary cryptography inside Internet of Things (IoT) space is discussed with a few patterns in the structure of lightweight strategies.

7.3 Optimal Data Hiding Model

Image DH approach has three essential things such as sender, receiver and finally the data hider. When the sender encrypts image by LWC encryption strategy, the data hider inserts the secret data with encrypted images in WSN. To select a DH key, the optimization technique called ECS-cuckoo eggs-based method is utilized. Using this procedure, the secret information is inserted within the original image. When the DH is finished, at that point, the extraction strategy is utilized to remove and decrypt the first image and secret data. The decrypted image, with great quality approximating to original image, can be acquired just with the encryption key for image security in WSN [27]. At this point, when the encryption key and DH key are accessible, the receiver can extricate the embedded information.

7.3.1 Digital Image Encryption

The architect of an encryption scheme should share the security keys, expected to recoup the first data, with proposed recipients so that other undesirable people can be prevented from accessing the information. The key encryption is slower when contrasted with private key encryption as a result of different keys distribution management between clients. It is mathematically expressed by

$$E_{i,j,k} = (DI_{i,j}/2^k)\mathrm{mod}2 \quad k = 1, 2 \ldots n \tag{7.1}$$

In order to ensure protection before presenting the image to data hider, the substance proprietor encrypts the image content with the help of a stream cipher route and hash function method.

7.3.2 Light Weight Hash Function

An ordinary hash work has a substantial interior state size and high power utilization, which may not be favorable for asset-compelled gadgets. This hash function has been actualized in the 4-bit S-box, and it satisfies the ciphers. For most of the part, the hash function has two critical procedures such as compression and collision models whereas Fig. 7.1 shows the general perspective of hash function. A perfect hash function must contain the properties of arbitrary oracle [25]. Despite the fact that there exists no irregular oracle, a hash function development should meet the security basis as it has some security properties.

- **Collision**: The function of collision is to find two different secret information, a1 and a0, so the hash function is $H(a0) = a1$. It requires $2^{n/2}$ process.
- **Pre image Resistance**: The hash values of $H(a)$ has a problem to find the original message and it require 2^n process.
- **Second Resistance**: The input message is difficult to find multi-input messages due to which the process is represented by $H(a0) = H(a1)$ whereas the hash function authentication includes 2^n process.

Traditional hash capacity is utilized to strengthen the contribution from huge size of 264 bits. In major objective rules of lightweight hash, the regular data measure is used. These hash functions, which are improved for short messages, may be increasingly sensible for lightweight image security process.

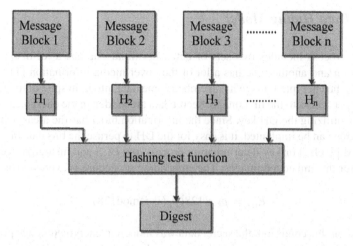

Fig. 7.1 General view of the hash Function

7.3.3 Encryption Model

Hash functions utilize a compression work that takes contribution from a few expressions of the affixing variable, symbolized by Hi, and a few expressions of the image extract, indicated by A. A single 64-bit chaining variable and the encryption are expressed by

$$H_i' = Enc(Hash_i, A) + H_{(i)} \qquad (7.2)$$

Based on the condition (7.1), the 256-pixel original images' matrix with lines and sections for which the scope of the pixel value is between 0 and 255 values. At this compression function, 64 bits of chaining variable and 80 bits of message-related information were compacted. Hash function assessment includes 128 bits in length whereas 64 bits of data were hashed and is processed by

$$H_1' = Enc(Hash_1, \| A) + H_1 \qquad (7.3)$$

In light of the above hash values, the images were encrypted by utilizing the public keys. This key was chosen by the prime number framework. It is known only to the sender who encrypted the information and should likewise be known to the receiver so as to recuperate the first plain image.

7.3.4 Data Hiding Model

The DH process includes two sets of data, the initial one has a lot of implanted information and another one has a lot of the cover media information [11]. When DH gains the encrypted image it can embed some data into it, in spite of the fact that it do not get access to the first image. Secret data are hidden in the encrypted original image by utilizing the DH key. Since the information hider has the areas, where the information can be implanted, it is easy for the DH to peruse the bits data in LSBs of encrypted pixels. For key determination process, the ECS and histogram were also created for the embedded images. The embedding procedure is expressed as

$$E_{i,j} = p_k * 128(p_e^{''}(i, j) \mathrm{mod} 128) \qquad (7.4)$$

Based on this condition, the secret data was hidden in encrypted image pixels p_e. Optimal DH key [28] can encrypt the to-be-embedded message. Along these lines, it is impossible to distinguish its essence after installing it in the marked encrypted image.

7.3.5 Optimal Data Hiding Key Selection

The most extreme security can be achieved for a given key size to take care of the issue of finding optimal key size for a given security level and to perform this, the EHS is utilized. From the optimal key, which was acquired by deliberately underestimating the computational exertion required for the security level, the target function of the DH key is

$$Fitness = \mathrm{max}(Throughput) \qquad (7.5)$$

This condition is to discover the throughput dependent on the optimal hiding keys, in all the implanted images. Another chance is that the value of the fitness function of the cuckoo egg is not exactly or arbitrarily equivalent to the fitness function value.

7.3.6 Background of ECS

Cuckoo is the best-known brood parasite. Some host-winged birds can connect straightforwardly with the encroaching cuckoo [29]. In the event that when host-winged birds distinguish its eggs from other birds' eggs, it either discards those eggs from its nest or just vacate and build another nest. The acquired solution is another solution dependent on the current one and the adjustment of a few attributes. In its simplest form, each home has one egg of cuckoo whereas other homes may have different eggs indicating a lot of solutions.

(1) **Levy Flight**

Levy flight is an arbitrary walk in which the steps are characterized with regards to the step-lengths that have a certain probability distribution, with the directions being random. This random walk can be observed in animals and insects. So the key solution is indicated by

$$hk_{n(new)}^i = hk_n^i + \beta * w * Levy(\delta) \tag{7.6}$$

$$\therefore Levy(\delta) = t(-\delta), \quad < \delta < 2 \tag{7.7}$$

The new solutions are developed and the step size was found with the improved method proposed in the current research. Each egg in home shows an answer and a cuckoo egg demonstrates another solution where the objective is to replace the weaker fitness with another result.

(2) **Enhanced updating procedure for data hiding levy selection**

At that point, the step size decides how far an irregular walker can go, for a fixed number of iterations. The effects for the last solution are estimated since it incites assorted new arrangements and is set to different characteristics. The condition for step size is

$$Stepsize = 0.01 \frac{dk^{t+1}}{|v^{t+1}|^{1/\beta}} (v - dk_{best}) \tag{7.8}$$

Here v as vector support to nest arrangement. This improved algorithm is based upon the best-found plan yet it is not reviewed in any different memories since the best solution, as demonstrated by the calculation, is continually held among the current courses of action.

(3) **Termination process**

Until getting the greatest throughput with optimal hiding key, the procedure is rehashed. Step size and levy flight-based are the bases for ECS functioning. The value of exchanging parameter was directly diminished as the ECS calculation was searching for worldwide least value of the test function.

Pseudo code for ECS
Initialization $(hk_1 hk_2, \ldots hk_n)$
Begin
Evaluate objective function (Eq. 7.5)
While (Iteration < maximum Generation)
 End
Update new keys by levy flight and Enhanced step size values
Again find (Eq. 7.5)
 Select the nest position among all position

If (Fitness (new) > fitness)
Replace the new solution
End if
Keep the best solutions and rank that solution
Freeze current best solutions
End

In view of the above methodology, the encrypted image was implanted with secret data with optimal DH keys. At long last, the DH haphazardly permutes the places of components of XE, and in this way, a potential assailant cannot extricate the embedded information.

7.3.7 Embedded Image Decryption Model

At the receiving side, the encrypted embedded image was decrypted by utilizing the LWE—hash function. At that point, the embedded bits were removed and the first image was recuperated. The receiver has the encryption key of the content proprietor and the DH key of the data hider. A conceivable answer for this case must ensure the independence between data extraction and image decryption. This process is represented by

$$H(Dec) = C \oplus \mathrm{mod}(dk_k + Cipher, 256) \oplus pr_{k1} \qquad (7.9)$$

Condition (7.9) demonstrates the embedded image encryption method. Here rand is an arbitrary number produced with the public key. It is used for the technique of decryption and the portrayal is not generally unclear, depending on the system used. This entire structure is graphically represented in Fig. 7.2. As per the DH plan, after image decryption, the secret bit installed in each image block ought to be separated, and the decrypted image ought to be completely recuperated to accomplish the reversibility.

7.3.8 Data Extraction and Image Reconstruction

In image decryption technique, the recipient do not know the secret bits installed in each block whether 0 or 1 [30]. The hidden data could be expelled to get the first image. By utilizing the optimal DH key, the secret data or information can be extricated from the decrypted image which depends on the least significant bit pixel values.

$$D_{i,j} = Enc(i, j) - 2 - rand(i, j)\mathrm{mod}128 \qquad (7.10)$$

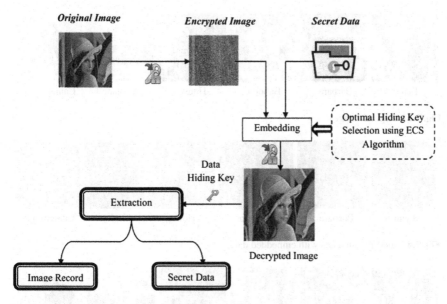

Fig. 7.2 Optimal key based data hiding model

Each block i.e., the first bit substituted by the inserted bit is zero which frames another block. At that point, the recipient accepts that the first bit is one and structures another block where one of them is the original block with the first bit. It is removed by

$$Ex = \begin{cases} 0 \; if \; EI^k \in C_l^k, C_q^k \\ 1 \; if \; EI^k \in \{C_l^k - 1, C_q^k + 1 \end{cases} \tag{7.11}$$

Here C_l and C_q are capacity factors and the content of secret data is preserved. In addition, he image blocks are shuffled in the encrypted image used, before performing the data embedding process. Finally, the extracted bits are concatenated to retrieve the additional message and the recovered blocks are collected to form the original image. From the secure image an attacker to separate the encryption since he should be time matched up with the system which might be troublesome for a malicious client.

7.4 Result and Analysis

In digital image security, the optimal key based DH was executed in MATLAB 2016 with i5 processor and 4GB RAM. The performance measures were PSNR, Embed-

| Lena | Barbara | Baboon | House | Airplane | Cameraman |

Fig. 7.3 Original images

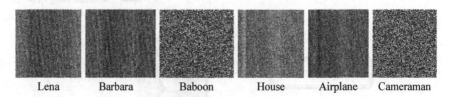

| Lena | Barbara | Baboon | House | Airplane | Cameraman |

Fig. 7.4 Encrypted images with embedded data

| Lena | Barbara | Baboon | House | Airplane | Cameraman |

Fig. 7.5 Decrypted images with extraction

ding Capacity (EC) (bpp) Throughput, DH capacity, Entropy, and most extreme capacity.

In Figs. (7.3, 7.4 and 7.5), the first image, then encrypted and decrypted with information are demonstrated. The embedding plan is reversible where the first cover substance can be consummately recouped in the wake of extracting the hidden data in WSN. In few situations, the encrypted image, containing the hidden data given by the server, should be decrypted by the approved client. After image encryption, the encrypted bits of every pixel are changed over into a gray value to produce an encrypted image (Figs. 7.6, 7.7).

Table 7.1 demonstrates the optimal DH with hash function consequences of all extensive test images. The PSNR of Lena image was 61.22 dB, Entropy and HC were 7.99 and 96.45% respectively at long last the objective function was 32.22. Not surprisingly, the normal lengths of three area maps and normal embedding limits got declined with expanding values. Here, the most extreme embedding capacity in one-layer embedding methodology was given. At that point, Fig. 7.3 demonstrates the similar examination of the proposed model and existing system i.e., DH-CS+ Hash function and DH+ Hash function. The PSNR values were moderately high in light of the fact that in this calculation the primary watermark was implanted in the encrypted image by just transforming one bit in each block of pixel values. While changing the EC, the PSNR got dissected by house image, for instance, the

Fig. 7.6 Comparative analysis for house image

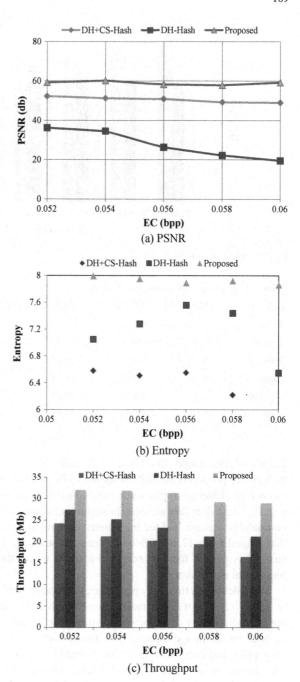

(a) PSNR

(b) Entropy

(c) Throughput

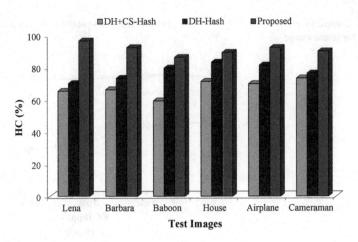

Fig. 7.7 HC comparison

Table 7.1 Experimental results of test images

Images	PSNR (dB)	EC (bpp)	Throughput (Mb)	Entropy	HC (%)
Lena	61.22	0.078	32.22	7.99	96.45
Barbara	59.52	0.088	28.22	7.79	92.33
Baboon	58.22	0.089	26.22	7.865	86.22
House	60.14	0.058	31.28	7.859	89.22
Airplane	59.85	0.077	29.55	7.88	92.22
Cameraman	56.18	0.086	27.85	7.93	90

EC was 0.056 (59.48 dB). Figure 7.3b demonstrates that the entropy values were high in the proposed method. If the EC differed, the entropy was diminishing and the stream cipher encryption was utilized in the proposed method i.e., 7.99. At long last, Fig. 7.3c demonstrates the throughput examination of all the images. Here, the comparable process had the maximum throughput of 0.058 in 32.22 DH procedure that ranged from 12 to 16% of the general strategy. In this manner, the proposed strategy is perfect for the encrypted images, where the essential controls are data hiding and extraction.

The calculation time for encryption, decryption, DH, and the extraction is shown in Table 7.2. This work, actualized on Matlab 2016a, can be observed usually that the unpredictability of the proposed strategy increases with the embedding rates are dictated by the image capacity. For airplane image, the time was 2.02 and 2.10 for encryption and decryption. Then the embedding and extraction were 0.76 and 0.92 for different images. The proposed technique distributed high intricacy on encryption as well as decryption steps with respect to DH procedure.

Table 7.2 Computational time (s) analysis

Images	Lena	Barbara	Baboon	House	House	Airplane	Cameraman
Encryption	1.15	1.88	1.29	1.27	1.89	2.02	1.67
Data embedding	0.78	0.79	0.65	0.88	0.86	0.76	0.77
Decryption	1.25	2.08	1.74	1.34	1.99	2.10	2.08
Data extracting	0.88	0.72	0.52	0.79	0.55	0.92	0.93
Total execution time	4.06	5.47	4.2	4.28	5.29	5.8	5.45

7.5 Conclusion

This section detailed data hiding with image security process. This work consolidated the data hiding and LWC hash function. The proposed algorithm permits the extraction of two embedded watermarks from the encrypted image and the careful image recovery after decryption and complete extraction of digital images in WSN. With regards to enhancing the security level of the work, the current research work utilized optimization (ECS) on data hiding key selection model. As per the data-hiding key, with the guide of objective work in digital image, the embedded data can be effectively extricated while the first image can be excellently recuperated. From the usage, the values of security measures such as PSNR, throughput, entropy, HC and ER were 59.188 dB, 0.079 bpp, 29.22 Mb, 7.88 and 91.073% respectively. As conventional encryption components might be unfeasible for WSN because of high overhead of processing and correspondence, a possible arrangement could be actually the mix of encryption algorithms. In future, progressive and extensive exertion are expected to decide the ideal parameters of the least noteworthy bit to expand the security level. Additionally the future studies are suggested to utilize attribute-based encryption or other stem ciphers.

References

1. Shankar, K., & Eswaran, P. (2016). RGB-based secure share creation in visual cryptography using optimal elliptic curve cryptography technique. *Journal of Circuits, Systems and Computers, 25*(11), 1650138.
2. Zhang, W., Ma, K., & Yu, N. (2014). Reversibility improved data hiding in encrypted images. *Signal Processing, 94,* 118–127.
3. Qin, C., & Zhang, X. (2015). Effective reversible data hiding in encrypted image with privacy protection for image content. *Journal of Visual Communication and Image Representation, 31,* 154–164.
4. Shankar, K., Elhoseny, M., Chelvi, E. D., Lakshmanaprabu, S. K., & Wu, W. (2018). An efficient optimal key based chaos function for medical image security. *IEEE Access, 6,* 77145–77154.
5. Al-Afandy, K. A., Faragallah, O. S., Elmhalawy, A., El-Rabaie, E. S. M., & El-Banby, G. M. (2016, October). High security data hiding using image cropping and LSB least significant

bit steganography. In *2016 4th IEEE International Colloquium on Information Science and Technology (CiSt)* (pp. 400–404). IEEE.

6. Ilayaraja, M., Shankar, K., & Devika, G. (2017). A modified symmetric key cryptography method for secure data transmission. *International Journal of Pure and Applied Mathematics, 116*(10), 301–308.

7. Mohamed, R. E., Ghanem, W. R., Khalil, A. T., Elhoseny, M., Sajjad, M., & Mohamed, M. A. (2018). Energy efficient collaborative proactive routing protocol for wireless sensor network. *Computer Networks*. Available online 2018, June 19.

8. Gupta, D., Khanna, A., Shankar, K., Furtado, V., & Rodrigues, J. J. Efficient artificial fish swarm based clustering approach on mobility aware energy-efficient for MANET. *Transactions on Emerging Telecommunications Technologies* (e3524). https://doi.org/10.1002/ett.3524.

9. Sreekumar, S., & Salam, V. (2014). Advanced reversible data hiding with encrypted data. *arXiv preprint* arXiv:1408.0733.

10. Hong, W., Chen, T. S., & Shiu, C. W. (2009). Reversible data hiding for high quality images using modification of prediction errors. *Journal of Systems and Software, 82*(11), 1833–1842.

11. Li, M., Fan, H., Ren, H., Lu, D., Xiao, D., & Li, Y. (2018). Meaningful image encryption based on reversible data hiding in compressive sensing domain. *Security and Communication Networks*.

12. Shankar, K., & Eswaran, P. (2017). RGB based multiple share creation in visual cryptography with aid of elliptic curve cryptography. *China Communications, 14*(2), 118–130.

13. Shankar, K., Lakshmanaprabu, S. K., Gupta, D., Khanna, A. & de Albuquerque, V. H. C. Adaptive optimal multi key based encryption for digital image security. *Concurrency and Computation: Practice and Experience* (e5122).

14. Elsayed, W., Elhoseny, M., Sabbeh, S., Riad, A. (2017). Self-maintenance model for wireless sensor networks. Computers and Electrical Engineering. Available online 2017, December (In Press).

15. Aminudin, N., Maseleno, A., Shankar K, Hemalatha, S., Sathesh kumar, K., Fauzi1, et al. Nur algorithm on data encryption and decryption. *International Journal of Engineering & Technology, 7*(2.26), 109–118.

16. Elhoseny, M., Elminir, H., Riad, A., & Yuan, X. (2016). A secure data routing schema for WSN using elliptic curve cryptography and homomorphic encryption. *Journal of King Saud University-Computer and Information Sciences, 28*(3), 262–275.

17. Elhoseny, M., Shankar, K., Lakshmanaprabu, S. K., Maseleno, A., & Arunkumar, N. (2018). Hybrid optimization with cryptography encryption for medical image security in Internet of Things. *Neural Computing and Applications*, 1–15.

18. Shankar, K., & Eswaran, P. (2016, January). A new k out of n secret image sharing scheme in visual cryptography. In *Intelligent Systems and Control (ISCO), 2016 10th International Conference* (pp. 1–6). IEEE.

19. Buchanan, W. J., Li, S., & Asif, R. (2017). Lightweight cryptography methods. *Journal of Cyber Security Technology, 1*(3–4), 187–201.

20. Zhang, X. (2011). Reversible data hiding in encrypted image. *IEEE Signal Processing Letters, 18*(4), 255–258.

21. Xu, D., Chen, K., Wang, R., & Su, S. (2018). Separable reversible data hiding in encrypted images based on two-dimensional histogram modification. *Security and Communication Networks*, 2018.

22. Yi, S., Zhou, Y., & Hua, Z. (2018). Reversible data hiding in encrypted images using adaptive block-level prediction-error expansion. *Signal Processing: Image Communication, 64*, 78–88.

23. Elhoseny, M., Yuan, X., ElMinir, H. K., & Riad, A. M. (2016). An energy efficient encryption method for secure dynamic WSN. *Security and Communication Networks, Wiley, 9*(13), 2024–2031.

24. Qin, C., Zhang, W., Cao, F., Zhang, X., & Chang, C. C. (2018). Separable reversible data hiding in encrypted images via adaptive embedding strategy with block selection. *Signal Processing, 153*, 109–122.

25. Hammad, B. T., Jamil, N., Rusli, M. E., & Zaba, M. R. A survey of lightweight cryptographic hash function.
26. Elhoseny, M., Tharwat, A., Farouk, A., & Hassanien, A. E. (2017). K-coverage model based on genetic algorithm to extend WSN lifetime. *IEEE Sensors Letters, 1*(4), 1–4. IEEE.
27. Elsayed, W., Elhoseny, M., Riad, A. M., & Hassanien, A. E. (2017). Autonomic self-healing approach to eliminate hardware faults in wireless sensor networks. In *3rd International Conference on Advanced Intelligent Systems and Informatics (AISI2017)* (2017, September 9–11). Cairo-Egypt: Springer.
28. Shankar, K., Elhoseny, M., Kumar, R. S., Lakshmanaprabu, S. K., & Yuan, X. (2018). Secret image sharing scheme with encrypted shadow images using optimal homomorphic encryption technique. *Journal of Ambient Intelligence and Humanized Computing*, 1–13. https://doi.org/10.1007/s12652-018-1161-0.
29. Majumdar, D., & Mallick, S. (2016, September). Cuckoo search algorithm for constraint satisfaction and optimization. In *Research in Computational Intelligence and Communication Networks (ICRCICN), 2016 Second International Conference* (pp. 235–240). IEEE.
30. Fedoseev, V. (2017). A model for data hiding system description. In *3rd International Conference on Information Technology and Nanotechnology* (pp. 65–71).

Chapter 8
Optimal Lightweight Encryption Based Secret Share Creation Scheme for Digital Images in Wireless Sensor Networks

Abstract Due to the advances in the computerized world, security has turned into an indivisible issue while transmitting the image. Secret Image Sharing (SIS) scheme is an updated cryptographic strategy which can be utilized to transmit a unique image from the sender to the receiver with incomparable privacy and secrecy in Wireless Sensor Network (WSN). Most traditional security techniques for popular systems, similar to the Internet, are not appropriate for WSN, requesting appropriate examination around there. It manages the procedure utilized to remodel the data among justifiable and deep structures through encryption and decryption strategies under the intensity of the keys. In this part, a high-security model for DI is proposed using a secret image share creation scheme with a novel Light Weight Symmetric Algorithm (LWSA). This cryptographic strategy partitioned the secret image into shares for the security analysis by optimal key selection in LWSA technique; here Adaptive Particle Swarm Optimization (APSO) is used. Moreover, this methodology is applying full encryption to image transmission over WSNs by offloading the computational outstanding task at hand of sensors to a server. The proposed LWSA plot offered better security for shares and furthermore reduced the deceitful shares of the secret image. The trial results and the investigations inferred that the proposed model can viably encrypt the image with quick execution speed and expanded PSNR value.

Keywords Images · WSN · Secret image sharing scheme · Optimal key · Light weight symmetric algorithm · Adaptive particle swarm optimization

8.1 Introduction

In the present situation, due to fast advancements in the field of Information Technology, security has turned out to be a pretty much serious issue. To transmit classified information, the web has turned into an essential source and safe information exchange is the real challenge here [1]. Cryptographic procedures allow information security at the cost of the classification and reduce the possibility of enemies in WSN [2], communication process [3]. It manages the system which is utilized to

© Springer Nature Switzerland AG 2019   115
K. Shankar and M. Elhoseny, *Secure Image Transmission in Wireless Sensor Network (WSN) Applications*, Lecture Notes in Electrical Engineering 564,
https://doi.org/10.1007/978-3-030-20816-5_8

revamp the information among reasonable and unimaginable structures by utilizing encryption and decryption techniques under the intensity of the keys [4]. The work of the cryptography is to cover information of an all-inclusive technique. In such manner, the credit goes to Shamir for propelling an all-around acclaimed strategy for secret sharing method [5]. Visual Secret Sharing (VSS) has awed the scholarly community using which various VSS applications got expanded, for example, picture encryption, visual verification, picture stowing away and advanced watermarking cryptography for secure information transmission in network modeling [6, 7]. The images transmission in WSNs presents significant test which raises issues identified with its portrayal, its stockpiling and its transmission [8]. After the extraction period, the grey share algorithm is connected on R part alone and n number of R grey shares is created. In the subsequent stages, the consolidated B and G segments with all-produced R dark offers to make shading shares [9]. The test controls various insignificant shares since all the shares must be required for different secrets to uncover. Accordingly, it is hard to control and utilize [10]. The generated shares are encoded by utilizing the Light Weight Encryption (LWE) display whereas the removed pixel values are utilized to make numerous offers (share 1, share 2... share n). The shares are separated into squares for the security [11]. These generated shares are transmitted to; sensor nodes have handling, storage and transmission constraints beginning from their asset compelled nature and the structure and task of remote visual sensor systems. The idleness of an encryption activity is the time between underlying solicitation for the encryption of a plain image and the answer that profits the related cipher image [12]. Here the arrangement of qualified shares is utilized to decode and to get the primary secret image. There is no key administration in this methodology since the key is not utilized [13]. The image information is mass measured and contains a ton of repetition which prompts difficulties in structuring vitality proficient picture transmission plots over WSN [14]. Symmetric key encryption utilizes the same key for both encryption and decryption of information. This encryption strategy is amazingly secure and generally quick [15]. A portion of the conventional symmetric key techniques is AES, DES, 3DES, BLOWFISH, IDEA, ECC, homomorphic encryption, and so forth [16]. Asymmetric key encryption utilizes two keys i.e., private and public key for correspondence between sender and the receiver. With the purpose of expanding security dimension of LWE, numerous optimization models are considered by researchers, for example, Genetic Algorithm (GA), Cuckoo Search (CS) algorithm, Particle Swarm Optimization (PSO) Gray Wolf Optimization (GWO) etc. [17].

8.2   Literature Review

The examination of Harn-Hsu's plan was not correct since their plan neglected to fulfill this element. In the event that one secret gets recreated, various unrecovered secrets can also be processed by any $t - 1$ investors in a misguided fashion. Another issue in Harn-Hsu's work was that the parameters were absurd and it is also discussed in the research study by Zhang et al. in 2018 [18]. The Harn-Hsu's plan had an error due to which another (t, n) multi-secret sharing plan was proposed and stretched out from Harn-Hsu's plan.

In 2018, Shankar et al. [19] recommended a wavelet-based secret image sharing plan with encrypted shadow images that utilizes ideal Homomorphic Encryption (*HE*) method. At first, the Discrete Wavelet Transform (DWT) is connected to the secret image to create subgroups. In this procedure, different shadows were made, encrypted and decrypted for each shadow. The encrypted shadow was recouped just by picking some subsets of these 'n' shadows that make straightforward and stack over one another. To enhance shadow security, each shadow was encrypted and decrypted by making use of the HE method.

Kanso and Ghebleh [20] proposed a plan that exploited the repetition in ordinary restorative images in order to diminish share sizes, and subsequently encourage storing and sharing. In this study, an altered run-length encoding method was utilized to cover the medicinal image. A broad execution examination was conducted for the proposed plan, incorporating a correlation with some current Shamir-type secret image sharing plans.

The idea of Visual Secret Sharing plan is to encode a secret image into 'n' irrational share images as per the study conducted by Shankar and Eswaran in 2017 [21]. They proposed an elliptic bend cryptography approach to increase the privacy and wellbeing of the image. The original-fangled method was used to create numerous offers which were exposed to encryption and decoding by methods for elliptic bend cryptography system.

In 2016, Shankar and Eswaran [22] proposed that a number of shares which are produced from secret images are counter-intuitive and contains certain message inside it, in visual cryptography. At the point, when all offers were heaped together, they generally uncover the secret of the image. The different offers were utilized to exchange the secret picture with the help of encryption and unscrambling process by ECC procedures. From the test outcomes, the PSNR value was found to be 65.73057, likewise, the mean square MSE esteem was 0.017367 and the connection CC was 1 for the unscrambled image with no mutilation in the first picture and the ideal PSNR.

In the Chaotic (C—work) process, the security was examined like dissemination just as uncertainty [23]. In light of the underlying conditions, diverse arbitrary numbers were created for each guide from chaotic maps. Adaptive Grasshopper Optimization (AGO) algorithm with PSNR and CC fitness work was proposed to pick the ideal secret and public keys of the framework among the arbitrary numbers.

A novel PSO for dynamic wireless sensor organize MOP to quicken information move in systems and decrease vitality misfortunes [24]. By and large, in DMOPs, the enhancement time frame is broken into a few equivalent sub periods. The method execution is demonstrated by applying it on three benchmark issues that were browsed the writing and one structure issue from the engineering area.

WSNs might be inclined to programming failure, unreliable remote associations, malevolent assaults, and equipment blames; that influence the system execution to debase essentially amid its life expectancy. One of these outstanding difficulties that influence the system execution is the adaptation to internal failure, it's proposed by

Elsayed et al. [25], the structure and difficulties of remote sensor systems and the primary ideas of self-recuperating for blame administration in WSN are discussed.

8.3 Motivation of This Study

Enhancing the protection and security consciousness has led to expanded research (and improvement) about the techniques that compute helpful data in a safe form. Data integration and sharing have been a long-standing challenge for the database wireless networks. Numerous researches were conducted to consolidate the encryption method, to decrease the time required to anchor the secret information, for quicker communication and better security for the secret information in WSN. The principle motivation behind this investigation is to plan a protected structure for defending digital images. This can be accomplished by decreasing the burdens found in the current techniques and by testing new methods created from the parts of existing solutions. The primary focal point of this research is to enhance the security of images by utilizing different optimization algorithms for a few stages. The proposed solution can be expanded with high accuracy rate compared to existing solutions.

8.4 Secret Image Share Creation Model

Visual Cryptography (VC) is an innovative idea, originally presented by Naor and Shamir in 1994. It is a strategy to encode a secret image into shares in an orderly fashion, such that the stacking of adequate number of shares uncovers the secret image. However, the initial image winds up as noticeable one to the exposed eye just by overlaying cipher transparencies known as 'shares' finished among the encryption procedure. In (S, n) VC model $(2 \leq S \leq n)$, the secret image is encoded into 'n' transparencies, called shares, just if at any rate when S shares are superimposed together, it can uncover the secret image. However, no information can be retrieved as much as S shares. In this plan, an image was broken into n shares so that somebody with all n shares could decrypt the image, while any $n - 1$ shares uncover no information about the initial image.

Pixel generation of images: The pixel estimations (P_v) of the secret color image (input) were extracted and taken as pixel values. These qualities are independently demonstrated as a matrix; the measure of the matrix was similar size to that of the first image matrix (a,b). Each pixel from the secret image was encoded into different subpixels in each shared image, utilizing a matrix, to decide the color of the pixels.

8.4.1 Secret Image Sharing Scheme for WSN Image Transmission

- SIS Scheme was utilized to make the shares from their pixel values. At first, an image of size (m × n) was selected; 'm' signifies the width and 'n' means the height of the image. The separated pixel values were utilized to make multiple shares (share 1, share 2… share n).
- Each share involved both highly contrasting pixels in the state of noise and was uncommonly extensive in measurement when contrasted with that of the secret image in WSN system. At that point, the matrices were made for each pixel, and furthermore, the variable was established, for example, image color and index value.
- Each and every share relies upon the pixel estimation of the cover image. The share for cover image is independently shown as C_i as follows

$$C_i = \int_1^S \lim_{S \to 1tnp} C_{ij} \tag{8.1}$$

$$2^i = S_n \tag{8.2}$$

- Different shares were created from the pixel values by utilizing this share creation scheme. The absolute number of shares was determined as the condition beneath. i indicates the basis matrix and S recognizes the number of shares shows in Eq. (8.2). Once the shares are generated, cryptographic algorithm are utilized to secure the shares in WSN transmission process.

8.4.2 Securing Images Using Light Weight Symmetric Cryptographic Algorithm

WSN image security, Lightweight cryptography is a subfield of cryptography that can provides customized arrangements to asset compelled tool. There has been a lot of work done by the scholarly network with regards to lightweight cryptography; this incorporates productive usage of customary cryptography principles, the structure and examination of new lightweight symmetric algorithms with APSO and protocols.

Light Weight Symmetric Algorithm (LWSA)
LWSA is a cryptographic cipher which is dependent on a lot of data security in network model, e especially for WSN security process. This encryption strategy utilizes what is known as a block cipher algorithm to guarantee that the images can be put away safely. The first phase of the current study security arrangement of the image is encryption. Here an encryption v is proposed to enhance the security of the information record.

(i) Let the input files, $X = \{S_1, S_2, \ldots S_n\}$ fed as input for encryption. From the input file, a position of dissimilar keys was extracted as $W = \{W_1, W_2, \ldots W_n\}$. Here, the created optimal key was utilized by the authorized client to recover documents from the cloud. The input information was decrypted utilizing the symmetric algorithm which enhances the security of the information put away by the information proprietor in the cloud.

(ii) The quantity of rounds in LWSA varies and relies upon the length of the key. LWSA utilized 10 rounds for bit keys, 12 rounds for 192-bit keys and 14 rounds for 256-bit keys. Every one of these rounds utilized an alternate 128-bit round key, which was determined from the first LWSA key. The key network was identified with the beneath table:

K0	K4	K8	K12
K1	K5	K9	K13
K2	K6	K10	K14
K3	K7	K11	K15

(iii) The operations of LWSA were sub-bytes: Shift row, Mix column and add the round key. The operation is visually shown in the Fig. 8.1.

Adaptive Particle Swarm Optimization (APSO) for Key selection

Particle Swarm Optimization (PSO) is a population-based stochastic optimization procedure created by Dr. Eberhart and Dr. Kennedy in 1995, enlivened by social behavior of birds rushing or fish tutoring. The framework is instated with a population of random arrangements and looks for optima by updating generations. In PSO, the potential solutions, otherwise called as particles, fly during the problem space by following the present ideal particles.

APSO algorithm procedure: The execution procedure includes five phases, initialization, fitness assessment, $Position_{best}$ and $Global_{best}$ computation, position and velocity updating and end criteria. The general procedure is shown in the flowchart beneath.

Step 1: Initialize the particles (key matrix) with random positions and their consequent velocities. Here, the intention of the proposed APSO is to generate establish on prime values.

$$I_i = I_1, I_2, \ldots I_n \qquad (8.3)$$

where, the value of i ranges from 1, 2, and 3...n.

Step 2: The fitness ability of the PSO algorithm is determined on the basis of objective work. Here, the commitment of optimization is to accomplish an ideal estimation of both FR and RR from the specified set of values.

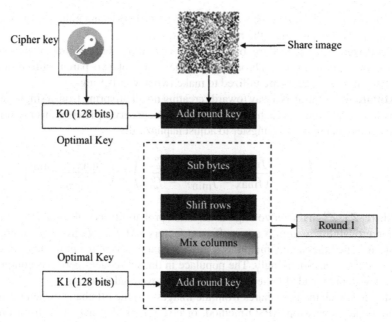

Fig. 8.1 Operations of LWSA

Step 3: Towards the beginning, the fitness value is decided nearly for every single particle. The ideal ones are chosen as $Global_{best}$ and $Position_{best}$ values among the fitness values. After the iteration, $Position_{best}$ is chosen as the current ideal fitness value whereas $Global_{best}$ is chosen as the general best fitness value. In the event that the present value is better, the contrast fitness estimation of particle and its $Position_{best}$, are set equivalent to the present value.

Step 4: The formulation for updating the velocity as well as the location of the particles in the inventive PSO are given as:

$$v_i(t+1) = v_i(t) + b_1 rand(Position_{best}(t) - r_i(t))$$
$$+ b_2 rand(Global_{best} - r_i(t)) \qquad (8.4)$$

$$r_i(t+1) = r_i(t) + v_i(t+1) \qquad (8.5)$$

where, V_i is the particle velocity, r_i is the current position of a particle, rand is a random number between (0, 1) and b_1, b_2 are learning factors, usually $b_1 = b_2 = 2$. As indicated by the updated methodology on (8.5), the *ith* particle position is coordinated by the situation of global best arrangement and position best arrangement. At that point, the fitness is discovered for another updated result. Here the result is refreshed on the basis of irregular estimation of position and the velocity. In view of this system, the researchers adjusted to pick the hybrid and mutation of Genetic Algorithm (GA) to select the best random value.

Crossover: The procedure of crossover is nothing but arbitrarily picking a crossover point inside a chromosome. The trades of two parent chromosomes between these focuses are utilized to make two new offspring.

Mutation: Mutation is a way towards creating novel offspring from a single parent by which the variety of each chromosome can be conserved. There is an opportunity to get the quality of a youngster to adjust haphazardly.

$$Z_d = z_d^0 + \left(1 + \varepsilon \frac{(f\text{max} - f\text{min})^\alpha - f_{avg}^\alpha}{\delta(f\text{max} - f\text{min})^\alpha - f_{avg}^\alpha}\right) \delta\left(\frac{f\text{max} - f\text{min}}{f_{avg}}\right)\alpha \quad (8.6)$$

where, Z_d, Z_d^0 characterize original mutation probability as well as adaptive mutation probability; α and δ are coefficient factors and f_{avg}, $f\text{max}$, $f\text{min}$ represent the average fitness, maximal fitness and minimal fitness of the individuals of each generation correspondingly. The populace in the subsequent period is comprised of new people made by the procedure of crossover and mutation.

Step 5: Check the new shares from the images. On the off chance when the ideal offers are done, stop APSO method, or else rehash the procedure from fitness assessment.

Optimal key are made for the safe image transmission and they keep up the image data privately. After that, the image shares are partitioned into squares. After this procedure, these offers are apportioned into blocks and the blocks of every part are encoded utilizing the cryptographic method.

Sub-bytes operation: The sub-bytes activity is a non-linear byte substitution that works on every byte of the state autonomously. The substitution table(S-Box) is invertible and is built using two changes. (a) The multiplicative reverse in Rijndael's limited field is taken (b) the relative change is applied and recorded in the Rijndael documentation.

Shift row operation: Each one of the four columns in the framework is moved to one side. Any entries that fall off' are re-embedded on the correct side of the column. The shift is done as pursue:

- The first row is not shifted.
- The second row is shifted one (byte) position to the left.
- The third row is shifted two positions to the left.
- The fourth row is shifted three positions to the left.

Mix column operation: Every segment of four bytes was changed utilizing an exceptional numerical task. This task takes information as four bytes of one segment and results would be four totally new bytes, which supplant the first segment. The outcome was another new network comprising of 16 new bytes. It must be noticed that this progression was not performed in the last round.

Add round key operation: The 16 bytes of the framework were considered as 128 bits and were XOR-ed to 128 bits of the round key. On the off chance, this is the last round, the result was the ciphered content. The subsequent 128 bits were translated as 16 bytes and another comparable round was started.

  (iv)   At long last, the encoded information and the separated keyword are put away in the cloud.

  (v)   The procedure of decryption of LWSA cipher content is similar to encryption procedure in inverted order. Each round comprises of four procedures directed in the turnaround request like (I) Add round key, (ii) Mix sections, (iii) Shift rows, and (iv) Byte substitution.

  (vi)   Since the sub-forms in each round are backward way, the encryption and decryption algorithms should be independently actualized, despite the fact that they are extremely and firmly related. The LWSA security is guaranteed just on the off chance that it is effectively executed and great key administration is utilized.

  (vii) Decryption is the opposite approach of encryption which speaks to the methodology of moving over the encrypted substance into one of its kind images. In this procedure, the optimal private key (K) is utilized to decode the message and the direct C11 is used toward decoding the pixel point.

  (viii) In the decryption strategy, the secured share images are recovered by adjusting the decrypted images back to their structures. In cryptography, the cipher images are passed on after encryption to the decryption framework with optimal public keys.

8.4.3   State of the Art

In this research, a privacy-preserving method is proposed for images with high security utilizing the SIS method for WSN communication process. It can be recognized by splitting the images in the cloud according to the secret share task. In this procedure, various shares were produced from one image; the routing model important routing protocols are used. In the share creation process, each share is independently created using the new SIS technique. Various shares are utilized to exchange the secret image through encryption and decryption processes by LWSA-APSO methods in WSN security, its shows in Fig. 8.2. The proposed plan offers better security for shares and furthermore reduces the fake shares of the secret image. This strategy comprises of encryption and decryption forms, the optimal public key, and the private key which are created by utilizing the prime number and the base purpose of the images. By utilizing this strategy, the primary image is shared safely and its data is kept secure and private.

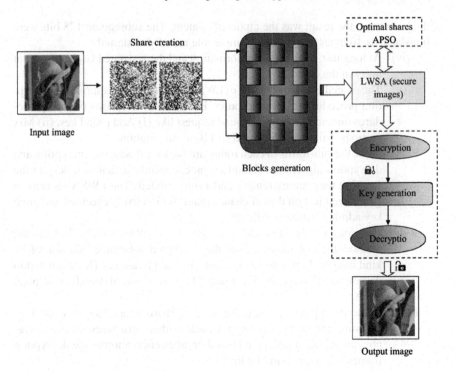

Fig. 8.2 Overview of securing methodology

8.5   Results and Analysis

In this area, the execution outcome of VSS image security in WSN system by various estimates was examined through various estimates such as PSNR, MSE, CC, and NPCR. This work was assessed or contrasted with other traditional techniques. Here the impressive images were Lena, Barbara, baboon, and house.

Figure 8.3 explains the PSNR analysis calculated with various algorithms. Here, for various iterations, the fitness value was calculated. At iteration 20, VSS + PSO algorithm obtained the fitness value 35, at 40, the fitness value obtained was 40. Likewise the iterations got increased and the fitness value also increased gradually. Moreover, VSS + APSO yielded the best fitness value when compared with other algorithms.

Table 8.1 explains various images for share creation and securing process. Here, various input image were taken, and for each image, the encrypted image, decrypted image, and reconstructed image are shown in Table 8.1. Table 8.2 explains the performance results for the proposed method. Lena, Barbara, baboon, and house were the images taken and for each image, the PSNR, CC, MSE and NPCR values were calculated. For each image, the different values were obtained for the proposed model.

Table 8.1 Image results for share creation and securing process

Table 8.2 Performance results for the proposed model

Images	PSNR	CC	MSE	NPCR
Lena	58.57	0.95	0.49	96.45
Barbara	61.48	0.99	0.29	92.22
Baboon	56.485	0.989	0.24	97.48
House	56.77	0.99	0.28	88.22

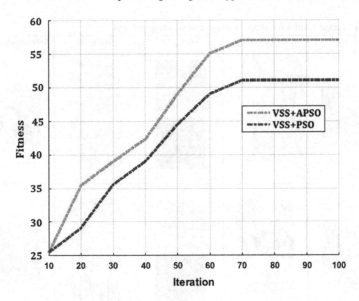

Fig. 8.3 Iteration versus fitness function

Figure 8.4 presents the time analysis for share creation, security and execution time for various images. For Lena image, the share creation time was 2.5, Barbara was 2.45, the baboon was 1.85 and house was 2.6. Then, the security for Lena image was 1.54, Barbara was 1.56, the baboon was 1.94 and the house was 2.3. The execution time is lengthy for all images.

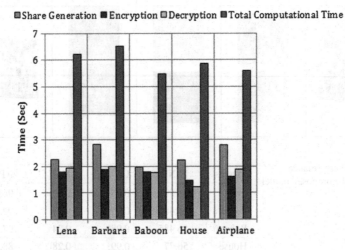

Fig. 8.4 Computational time analysis

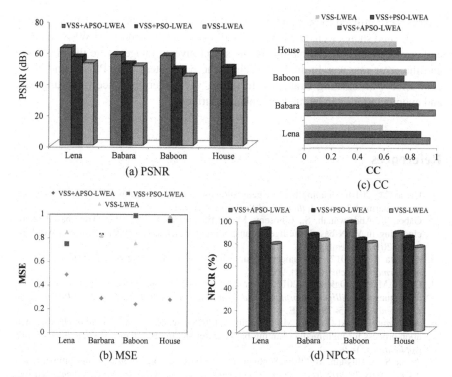

Fig. 8.5 Comparative analysis

Figure 8.5 presents the comparative analysis with various algorithms and various images. It was performed with various performance measures such as PSNR, MSE, CC and NPCR. In Fig. 8.5a, PSNR varied from 0 to 80. For all images, VSS-LWSA yielded better performance compared to other algorithms. In Fig. 8.5b, MSE varied from 0 to 1 and when compared with other algorithms, VSS-LWSA provided the least error. Similarly, for other measures, VSS-LWSA yielded better performance with various images.

8.6 Conclusion

A novel method for secure, privacy-preserving image storage and sharing the image data is proposed in the current chapter. An effective VSS model was utilized to share the secret images before encryption. To increase higher energy efficiency in WSN image transmission and broaden arrange lifetime. In this investigation a remote media arrange is viewed as which sensor nodes. In the VSS procedure, diverse images were utilized to assess the execution of the proposed strategy. In this procedure, diverse images were considered, for example, Lena, house, peppers, and monkey images were utilized to part and produce the shares. In this share creation process, the generated shares were anchored depending on (APSO) optimal-LWSA. Without any errors, the PSNR estimations of the secret diverse images were the maximum and the error rate was limited for each of the parameters (CC, PSNR, and NPR) contrasted with existing strategies. In future, yet numerous conceivable upgrades and augmentations can be made to enhance further.

References

1. Ulutas, M. (2010). Meaningful share generation for an increased number of secrets in the visual secret-sharing scheme. *Mathematical Problems in Engineering*.
2. Gaber, T., Abdelwahab, S., Elhoseny, M., & Hassanien, A. E. (2018). Trust-based secure clustering in WSN-based intelligent transportation systems. *Computer Networks*. Available online 2018, September 17.
3. Abusitta, A. H. (2012). A visualcryptography based digital image copyright protection. *Journal of Information Security, 3*(02), 96.
4. Ogiela, M. R., & Ogiela, L. (2015, November). Bio-inspired approaches for secret data sharing techniques. In *Intelligent Informatics and Biomedical Sciences (ICIIBMS), 2015 International Conference* (pp. 75–78). IEEE.
5. Al-Ghamdi, M., Al-Ghamdi, M., & Gutub, A. (2018). Security enhancement of shares generation process for multimedia counting-based secret-sharing technique. *Multimedia Tools and Applications*, 1–28.
6. Arora, S., & Hussain, M. (2018, September). Secure session key sharing using symmetric key cryptography. In *2018 International Conference on Advances in Computing, Communications and Informatics (ICACCI)* (pp. 850–855). IEEE.
7. Shankar, K., Devika, G., & Ilayaraja, M. (2017). Secure and efficient multi-secret image sharing scheme based on boolean operations and elliptic curve cryptography. *International Journal of Pure and Applied Mathematics, 116*(10), 293–300.
8. Elhoseny, M., Tharwat, A., Farouk, A., & Hassanien, A. E. (2017). K-coverage model based on genetic algorithm to extend WSN lifetime. *IEEE Sensors Letters, 1*(4), 1–4. IEEE.
9. Hodeish, M. E., Bukauskas, L., & Humbe, V. T. (2016). An Optimal (k, n) visual secret sharing scheme for information security. *Procedia Computer Science, 93*, 760–767.
10. Ulutas, M., Yazici, R., Nabiyev, V. V., & Ulutas, G. (2008, October). (2, 2)-Secret sharing scheme with improved share randomness. In *Computer and Information Sciences, 2008. ISCIS'08. 23rd International Symposium* (pp. 1–5). IEEE.
11. Shankar, K., & Eswaran, P. (2015). Sharing a secret image with encapsulated shares in visual cryptography. *Procedia Computer Science, 70*, 462–468.
12. Yan, X., Lu, Y., & Liu, L. (2019). A general progressive secret image sharing construction method. *Signal Processing: Image Communication, 71*, 66–75.
13. Shankar, K., & Eswaran, P. (2016, January). A new k out of n secret image sharing scheme in visual cryptography. In *Intelligent Systems and Control (ISCO), 2016 10th International Conference* (pp. 1–6). IEEE.
14. Elhoseny, M., Yuan, X., ElMinir, H. K., & Riad, A. M. (2016). An energy efficient encryption method for secure dynamic WSN. *Security and Communication Networks, 9*(13), 2024–2031. Wiley.
15. Patel, T., & Srivastava, R. (2016, August). A new technique for color share generation using visual cryptography. In *Inventive Computation Technologies (ICICT), International Conference* (Vol. 2, pp. 1–4). IEEE.
16. Shankar, K., & Eswaran, P. (2015). A secure visual secret share (VSS) creation scheme in visual cryptography using elliptic curve cryptography with optimization technique. *Australian Journal of Basic & Applied Science, 9*(36), 150–163.

17. Elhoseny, M., Shankar, K., Lakshmanaprabu, S. K., Maseleno, A., & Arunkumar, N. (2018). Hybrid optimization with cryptography encryption for medical image security in Internet of Things. *Neural Computing and Applications*, 1–15.
18. Zhang, T., Ke, X., & Liu, Y. (2018). (t, n) multi-secret sharing scheme extended from Harn-Hsu's scheme. *EURASIP Journal on Wireless Communications and Networking, 2018*(1), 71.
19. Shankar, K., Elhoseny, M., Kumar, R. S., Lakshmanaprabu, S. K., & Yuan, X. (2018). Secret image sharing scheme with encrypted shadow images using optimal homomorphic encryption technique. *Journal of Ambient Intelligence and Humanized Computing*, 1–13.
20. Kanso, A., & Ghebleh, M. (2018). An efficient lossless secret sharing scheme for medical images. *Journal of Visual Communication and Image Representation, 56*, 245–255.
21. Shankar, K., & Eswaran, P. (2017). RGB based multiple share creation in visual cryptography with aid of elliptic curve cryptography. *China Communications, 14*(2), 118–130.
22. Shankar, K., & Eswaran, P. (2016). RGB-based secure share creation in visual cryptography using optimal elliptic curve cryptography technique. *Journal of Circuits, Systems and Computers, 25*(11), 1650138.
23. Shankar, K., Elhoseny, M., Chelvi, E. D., Lakshmanaprabu, S. K., & Wu, W. (2018). An efficient optimal key based chaos function for medical image security. *IEEE Access, 6*, 77145–77154.
24. El-Shorbagy, M. A., Elhoseny, M., Hassanien, A. E., & Ahmed, S. H. (2018). A novel PSO algorithm for dynamic wireless sensor network multiobjective optimization problem. *Transactions on Emerging Telecommunications Technologies* (e3523).
25. Elsayed, W., Elhoseny, M., Riad, A. M., & Hassanien, A. E. (2017). Autonomic self-healing approach to eliminate hardware faults in wireless sensor networks. In *3rd International Conference on Advanced Intelligent Systems and Informatics (AISI2017)* (2017, September 9–11). Cairo-Egypt: Springer.

Chapter 9
Multiple Share Creation with Optimal Hash Function for Image Security in WSN Aid of OGWO

Abstract Visual share creation, with a security model, verifies the secret data transmitted by the sender. It uses human vision to decrypt the encrypted images without complex algorithms for secure communication in networking. The image sensing process joined with preparing force and wireless communication makes it rewarding for being abused in wealth in future. The proposed security module beats the downsides of the complex usage in traditional or ordinary cryptography models in wireless sensor network (WSN). This chapter proposed an optimal private key and public key-based hash function for secret share security modeling. Initially, the original Digital Images were considered to produce 'n' number of shares for security. When the quantity of share was expanded, the security level also got expanded. In the current technique, three shares were produced by visual secret share creation strategy when the shares were made; hash function and qualities were applied to secure the shares in images. For an optimal key selection model, opposition-based Grey Wolf Optimization (OGWO) was considered. Based on this ideal private key and public keys, the share images were encrypted and decrypted between the sender and the receiver. The test results demonstrated better security, less computational time and the most extreme entropy with high PSNR values when compared with other techniques.

Keywords Secret share creation · WSN · Optimization · Hash function · Light weight encryption · Oppositional process

9.1 Introduction

The security suspicions of multimedia data is a testing problem as the measure of the information is immense though the processing of information consequently in a continuous manner is mandatory too [1]. The traditional strategy to ensure security is by encrypting the data utilizing block encryption techniques, for example, Advanced Encryption Standard (AES) in WSN [2]. Consequently, it is important to focus on the security of multimedia data. Since the image pixels are progressively associated, the conventional encryption plans are not effective [3].

© Springer Nature Switzerland AG 2019   131
K. Shankar and M. Elhoseny, *Secure Image Transmission in Wireless Sensor Network*
(WSN) Applications, Lecture Notes in Electrical Engineering 564,
https://doi.org/10.1007/978-3-030-20816-5_9

As a rule, sensor nodes have handling, storage and transmission limitations beginning from their resource constrained nature in WSN security [4]. Color Visual Cryptography is a developing field which encrypts the shading secret messages into different quantities of shading halftone image shares. Visual information pixel synchronization and error diffusion system empower the encryption of visual data with quality [5]. Synchronization translates the spot of pixels alongside the secret images and amid blunder dispersion, the shares are produced which is visible to the human visual framework [6].

Visual secret sharing technique was proposed by Naor and Shamir depending on the possibility that a gathering of k out of n participants can recreate the secret image [7]. The strategy encodes a secret image into n share or shadow images. Confidential information can be implanted in various edges since round shares are utilized in this strategy. However, an imperative challenge in these plans is that the first share is not made completely irregular [8]. There are numerous uses of VC such as incorporation of copyright security, general access structure, watermarking and visual confirmation [9]. The encryption strategies require the utilization of a secret key. Exposure of cryptographic key is likewise hazardous, as a solitary individual, cannot depend on the security of the key [10]. This hash function has been actualized in the 4-bit S-box, and it satisfies the PRESENT structure criteria. Sponge has 13 variations for different dimensions of crash/(second) preimage conflict and execution imperative [11]. A perfect hash function must contain the properties of the irregular prophet. Despite the fact random oracle does not exist, a hash function development should meet the security standards in network security for images [12]. The sensors conveyed to recover image previews and video streams will commonly request a greater number of assets than conventional scalar sensors in image transmission system in WSN system [13]. The security of sponge put together the structures depend with respect to their capacity. It is important to note that Lightweight Encryption (LWE) ought not to be compared with frail cryptography. Lightweight designs are essentially more fragile than traditional algorithms [14].

9.2   Related Works

Shankar and Eswaran [15] proposed Visual Cryptography (VC) as a strategy that protects a secret image, where the image is encoded into numerous shares and dispensed to different members. In the share creation process, the authors indicated a new condition for arbitrary frameworks after which XOR activities were performed to produce 'n' transparencies. In 2015, Bhadravati et al. [16] recommended a Scalable Secret Image Sharing (SSIS) technique that furnished slow recreation with smooth adaptability. Moreover, this strategy was stretched to recordings and a versatile secret video sharing (SSVS) technique was proposed. These two techniques are intended to compress multimedia. A novel PSO for dynamic WSN in MOP quicken information move in systems and decrease energy losses by El-Shorbagy et al. [17]. By and large, in DMOPs, the enhancement time frame is broken into a few equivalent subperiods.

Foe models are necessary to the structure of proven and secure cryptographic plans or protocols as opined by Do et al. in 2018 [18]. The classification plan for basic application-based foes utilized the stable security and the key papers were classified as per the proposed plan. At last, the ongoing work examines the models in the contemporary research field of IoT. Wavelet-based secret image sharing plan was proposed with encoded shadow images utilizing ideal Homomorphic Encryption (HE) system. At first, Discrete Wavelet Transform (DWT) was connected by Shankar et al. [19] to the secret image to create subbands. The encrypted shadow can be recuperated by just choosing some subsets of these 'n' shadows that makes it straightforward and stack over one another. To enhance shadow security, each shadow was encrypted and decrypted utilizing the HE procedure. In terms of worries regarding image quality, a new Opposition-based Harmony Search (OHS) algorithm was used to create the optimal key [20].

9.3 Contribution of the Work

In this chapter, the researchers proposed the cryptographic security in WSN investigation which is partitioned as two modules such as (i) Generation of "n" shares and (ii) Optimal key-based lightweight cryptography. This optimization-based encryption and decryption of digital image OGWO system is used to locate the private key of encryption model, for the routing process LEACH protocol used. The purpose behind choosing OGWO is to deliver a better candidate solution among the first and oppositional solutions for nearby ideal fitness function. This Lightweight Cryptographic Encryption and decryption performance are analyzed by random generation of public keys or private keys when compared with the regular cryptography model. The enhanced opposition learning is used to enhance the solution from the usage of security measures in order to provide better outcomes in terms of Maximum PSNR, least MAE.

9.3.1 Implementation Procedure of Grey Wolf Optimization (GWO)

GWO is inspired by the hunting pattern of grey wolves in wild life. Wolves live in a pack and are isolated into two dark wolves group such as male and female for dealing alternate wolves in the pack [21]. A solid social defeat pecking order is present in each pack. The hierarchical patterns of the grey wolves are developed by pack leaders. The leaders that have a place with the pack are sorted into four structures such as Alpha, Beta, Delta, and Omega which are shown in Fig. 9.1. The new solution updating method of GWO is separated into three stages i.e., prey searching, prey encirclement, and hunting behavior.

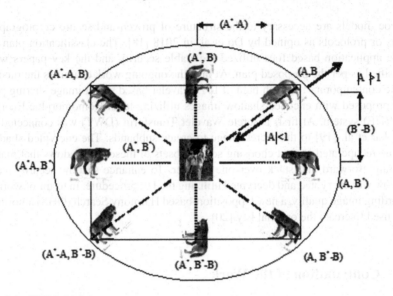

Fig. 9.1 GWO behavior

(i) **Oppositional model**

When moving towards the OBL heuristics, the opposite point, and the opposite number assume an important job. These two definitions are assessed through opposition-based populace initialization. The opposition solution generates real numbers which are considered by a, b and the opposition number is 'o' which are scientifically portrayed by

$$\hat{o} = a + b - o \tag{9.1}$$

The calculation of point in addition to the opposite point is performed at one go with the best solution. The calculation is constantly performed through all the present populations and for every solution, it is not performed alone for the initialization of the solution.

(ii) **New grey updating model**

The general initialization of grey wolves, within the search space, makes it ready for the search task. In order to look for the prey, the grey wolf starts preoccupation with one wolf, then onto the next and the wolves converge once the prey is detected. The encirclement is influenced after the detection of prey solution.

$$\vec{Q} = |A * \cdot s_a(i) - a(i)| \tag{9.2}$$

Here "i" denotes the present iteration whereas the coefficient vectors are $\vec{Q}$, A. The term is used to balance the separation between prey and the related grey wolves. The interruptions in the chasing ways of grey wolves are indicated as s. The position vector of the grey wolf is signified as a and the position.

After the completion of the prey inclusion method, the grey wolves focus on the pursuing (hunting) of prey [21]. The three hidden best solitary qualities accomplished, starting at as of recently, require the other intrigue specialists to change their conditions as per the situation of the best solution.

$$Q^{\alpha} = |A_1 \cdot s_{\alpha} - a|, \; Q^{\beta} = |A_2 \cdot s_{\beta} - a|, \; Q^{\delta} = |A_3 \cdot s_{\delta} - a| \qquad (9.3)$$

The new grey wolfs get refreshed for the SVD transform process. With diminishing A, half of the iterations are fixated on the examination $(A| < 1)$ and the other half is focused on the use. The GWO algorithm is dependent on the positions of α, β, and γ since it permits the search specialists for prey attack by updating their locations (positions). This enables it a ready-to-achieve optimal solution.

9.3.2 Cryptography Securing Model in WSN

The visual cryptographic scheme use the image shares as the decryption factor and due to this reality, the final output images are much clear to the human vision. Every pixel of the secret (unique) image, communicated as "m" upgraded variant, is known as shares for WSN image security. Again the RGB image is spoken as a group of sub-pixels and these sub-pixels form a share. The created image is encoded into 'm' various irrational shares that make use of secret sharing method. This created share is viewed as a contribution to the encryption and decryption models. The optimal key is chosen by previously-mentioned OGWO technique. When the images are encrypted by an optimal key, at that point, the decryption task is performed in the receiver side. Shares can be generated either with a block cipher or stream ciphers when the latter is favored, because of its superior exhibitions and encryption speed.

(1) Share creation

At first, the pixel estimations of the image are read to separate the RGB bands after which the shares of various bands are produced. Each share is an accumulation of sub-pixels, i.e. Share 1, 2 … n for R band. The other band share is also created which is defined as follows

$$Sh_p = \int_1^k \lim_{k \to 1tnp} I_{ab} \qquad (9.4)$$

where $a \; and \; b$ are the situations in the matrix whereas the separated bands and pixel measures are separated from the first images which are built into individual matrix

Fig. 9.2 Share creation

[22] and the share creation model is shown in Fig. 9.2. An inventive visual secret share generation procedure has been utilized to deliver two shares from each one of image in transmission process.

The steps of share creation is explained below:

(i)	Read the image Size height × width of the image.
(ii)	Generate three matrices of the original image for share 1, share 2 and share 3.
(iii)	Set the index factor as = 1
	Fix the width and height of considered matrix as ′1′
(iv)	Fix the index values
	If
	Index = 1
	Set Share 1 = Image Color
	Set Share 2 = Image size of particular matrix
	Set share 3 = Empty matrix
	Else
	Set Share 1 = Empty matrix
	Set Share 2 = Image color
	Set share 3 = Image size of particular matrix
	Else if
	Set Share 1 = Image size of the particular matrix
	Set Share 2 = Image color
	Set share 3 = Empty matrix
(v)	Based on the generated shares, create the blocks by the size of 4*4, finally the blocks are secured using the keys.
	$SB_1 = Share_{(1,2,3)} \oplus optimal \, key$ (9.5)

The aim of the secret offering plan is to encrypt a secret image into irrelevant offer images. It cannot discharge any information about the underlying image except when all the shares are procured. The created shares utilize XOR activity that is shown in this condition (9.5). Then the blocked shares are considered for LWE-based security analysis.

(2) Optimal key-based encryption and decryption

Taking the metrics into consideration as both in hardware case and software case, the most lightweight algorithms intend to utilize few internal states, short block, and key sizes. The selected asymmetric cryptography system hash function is utilized with ideal selection. A standard method, to construct a hash function, is to iteratively apply a fixed-input length compression function. The security level of the proposed hash functions can be expected against this technique. The initial step is to make the public key to encode the message and the public key is normally acquired from the recipient side. The second step is to make the private key on the recipient side for decryption of the secretly-shared image.

9.3.3 Hash Function

Hash functions are basically many-to-one functions while they map subjective length inputs to fixed length outputs and the input is normally bigger than the output. In a number of applications, it is used to end the troubled hash value than finding a crash between two self-assertive messages. Along these lines, the application, at which the hash function is being utilized, decides the security properties that it should accumulate [23].

Encryption
In this procedure, the isolated color bands of the first image were divided into blocks and each block was encrypted by making use of the encryption technique. For this thought about the blocked shares, the scope of pixel esteem was between 0 and 255 values and is communicated by

$$\text{hash} \, (\text{En}(\text{share}, \text{public}_{\text{optimal}-k1}), h_k) = h(\text{image}, h_k)$$

$$\text{hash}(\text{En}(\text{share}, \text{public}_{\text{optimal}-k2}), H_k) = h(\text{En}(\text{Image}, \text{public}_{\text{optimal}-k2}), h_k) \quad (9.6)$$

With the assistance of the condition (9.6), this initially divided the image into blocks; every measurement of request was 16×16. Towards separating, a total of 256 such blocks were attained. The algorithm utilized different midpoints while encoding distinctive information of images to even with a similar sequence dependent on hash function.

Decryption
In image decryption, the ideal private key was utilized to decrypt the message and the pixel point was decrypted depending on the bands. The security of the cipher should depend on the decryption keys pr_{k1}, pr_{k2}. Since an adversary can enhance the plain image from the observed cipher image it is communicated by

$$\text{Decrypted} = \text{Decrypt (Cipher share)} \qquad (9.7)$$

$$h(\text{Dec}) = \text{Ciphershare} \oplus \mod(h_k + \text{Cipher}, 256) \oplus \text{optimal}(pr_{k1}, pr_{k2}) \qquad (9.8)$$

The decryption process was directed by somewhere around one cryptographic key. In general, the keys used for the technique of decryption and portrayal are not generally indistinct but contingent upon the system used. The procedure was repeated until the most ideal key was achieved. The fitness measure was evaluated and it was checked with the fitness proportion of other old solutions; at that point the optimal key was acquired from the decryption procedure.

Optimal key selection procedure in WSN security

Initialize the opposition solutions
Generate initial population for key selection
Find the fitness model Fi = Max (Entropy)
Choose the best individual from the defined opposition and general solutions.
If
Condition satisfied?
                    End
Else
Update New Solutions
Prey Encirclement, hunting Behaviors
Again find the fitness
If (Optimal New key) > f (old key)
Store the best Solutions
iter = iter + 1
Terminate until getting the optimal; key with maximum fitness reached.

9.3.4 Arranged Image

According to the above technique, the shares were encrypted and stacked with these every single expanded block. The pixel estimations of three secret images were revealed since the hamming distance among these two extended blocks was 0. By then, the characteristics of the subsequent pixel for two stacked pixels were as following; dark, if both are black white and if both are white black generally. At long last, the OGWO got updated in encryption as well as decryption and then the resultant image was acquired.

9.4 Results and Analysis

In this section, the results about optimal hash function-based secret share security model was discussed using different measures like PSNR, Entropy, MAE, and NC, It was implemented in MATLAB 2016a with the system configuration, i5 processors with 4GB RAM. Moreover the proposed work is compared with other techniques like hash function with GWO and 'only hashes' function.

Table 9.1 depicts the proposed image security results. Here images such as Leena, Barbara, Baboon, house, and airplane were taken. These proposed results were attained based on the performance parameters such as PSNR, Entropy, MAE, and NC. Initially, the input image was taken and made shares for those images. Then for security purpose, the images were encrypted based on lightweight cryptography and the images were decrypted by optimizing both public and private keys in secure image transmission in WSN. Finally, the output image was attained with high PSNR, optimal entropy, minimum error, and optimal NC. The highest PSNR was achieved in the range of 61.27, optimal fitness was 7.99, MAE was 0.16, and NC equal to 1.

Figure 9.3 shows the computational time analyses for the proposed study. Here, the computational time for every image (Lena, Barbara, baboon, house, and airplane) was analyzed. The bar graph denotes the processing time for share generation, encryption, decryption, and computational time. More time was consumed during the share generation process. Then the encryption process took high processing time than the decryption process. Figure 9.4 shows the comparative analysis based on all performance parameters. In the comparison, the analysis demonstrated that the hash functions required some optimization to achieve the optimal value. So, the OGWO algorithm was used to optimize the key function. The hash function, with OGWO, achieved optimal performance when compared with separate GWO.

Table 9.2 shows the results of images with respect to two attacks (salt and pepper, and brute search attack). The performance parameters attained were low after the attack was formed in those images.

Fig. 9.3 Computational time analysis (s)

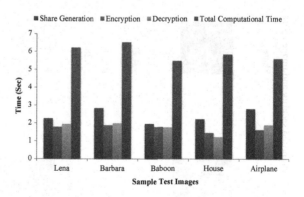

Table 9.1 Proposed security and image results

Image	Shares	Encrypted	Decrypted	Stacked
(a) *Lena*				
	PSNR	Entropy	MAE	NC
	59.28	7.88	0.29	0.98
(b) *Barbara*				
	PSNR	Entropy	MAE	NC
	61.271	7.74	0.16	0.99

(continued)

Table 9.1 (continued)

Image	Shares		Encrypted		Decrypted		Stacked
(c) *Baboon*							
	PSNR		Entropy		MAE		NC
	53.22		7.95		0.33		0.95
(d) *House*							
	PSNR		Entropy		MAE		NC
	58.74		7.89		0.62		1

(continued)

Table 9.1 (continued)

Image	Shares	Encrypted	Decrypted	Stacked
(e) *Airplane*				
	PSNR	Entropy	MAE	NC
	54.18	7.99	0.52	1

Fig. 9.4 Comparative analysis

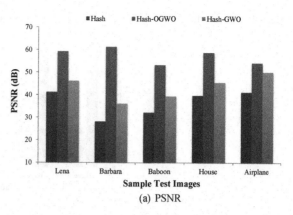

(a) PSNR

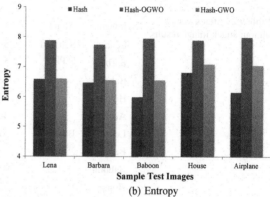

(b) Entropy

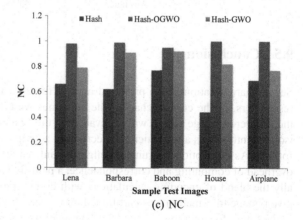

(c) NC

Fig. 9.4 (continued)

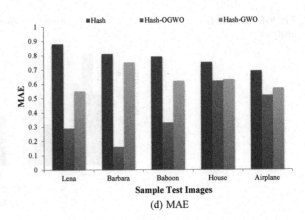

(d) MAE

Table 9.2 Attack versus without attack image results

Images	Attack	PSNR	Entropy	MSE	NC
Lena	Salt and pepper	53.22	6.95	0.82	0.75
Barbara		56.11	6.78	0.72	0.82
Baboon		48.72	7.02	0.62	0.81
House		39.89	6.85	1.02	0.76
Airplane		46.22	7.05	0.88	0.72
Lena	Brute search	51.09	6.44	0.49	0.66
Barbara		53.71	6.89	0.92	0.80
Baboon		43.18	7.64	1.07	0.76
House		32.22	5.58	0.66	0.61
Airplane		43.58	7.28	0.67	0.63

9.5 Conclusion

Secret share creation is the preferred area of research interest among the current researchers. In the current chapter, the outcomes were better than the measure of the recovered image which was equivalent to the first. The first secret image was isolated into shares after which the decryption activity of qualified shares was performed. As encryption instruments might be unfeasible for WSN because of high overhead of computing and correspondence, a plausible arrangement could be actually the blend of encoding calculations with cryptography in image security. The given image information was partitioned into secret sharing parts and were utilized in the future for Encryption. The OGWO model was utilized to select the ideal private and public keys of LWE-hash function; it is superior and progressively powerful enough in calculation purposes for image encryption. The proposed model enhanced the security level with least computational time and exhibited multifaceted nature of share creation in encryption and decryption models in WSN security. The maximum

PSNR was 61.22 dB and target function entropy was 7.99 in ideal hash function security. For further parts or research studies, the proposed model can be stretched out to imaginative share creation and quality-based encryption model.

References

1. Shankar, K., & Eswaran, P. (2015). Sharing a secret image with encapsulated shares in visual cryptography. *Procedia Computer Science, 70*, 462–468.
2. Ulutas, M., Yazici, R., Nabiyev, V. V., & Ulutas, G. (2008, October). (2, 2)-secret sharing scheme with improved share randomness. In *23rd International Symposium on Computer and Information Sciences, 2008, ISCIS'08* (pp. 1–5). IEEE.
3. Eswaran, P., & Shankar, K. (2017). Multi secret image sharing scheme based on DNA Cryptography with XOR. *International Journal of Pure and Applied Mathematics, 118*(7), 393–398.
4. Mohamed, R. E., Ghanem, W. R., Khalil, A. T., Elhoseny, M., Sajjad, M., & Mohamed, M. A. (2018). Energy efficient collaborative proactive routing protocol for wireless sensor network. *Computer Networks*. Available online June 19, 2018.
5. Shankar, K., & Eswaran, P. (2015). ECC based image encryption scheme with aid of optimization technique using differential evolution algorithm. *International Journal of Applied Engineering Research, 10*(55), 1841–1845.
6. Shankar, K., Devika, G., & Ilayaraja, M. (2017). Secure and efficient multi-secret image sharing scheme based on boolean operations and elliptic curve cryptography. *International Journal of Pure and Applied Mathematics, 116*(10), 293–300.
7. Parakh, A., & Kak, S. (2011). Space efficient secret sharing for implicit data security. *Information Sciences, 181*(2), 335–341.
8. Elhoseny, M., Yuan, X., El-Minir, H. K., & Riad, A. M. (2016). An energy efficient encryption method for secure dynamic WSN. *Security and Communication Networks, 9*(13), 2024–2031.
9. Cortier, V., & Steel, G. (2009, September). A generic security API for symmetric key management on cryptographic devices. In *European Symposium on Research in Computer Security* (pp. 605–620). Berlin, Heidelberg: Springer.
10. Elhoseny, M., Shankar, K., Lakshmanaprabu, S. K., Maseleno, A., & Arunkumar, N. (2018). Hybrid optimization with cryptography encryption for medical image security in Internet of Things. *Neural Computing and Applications*, 1–15.
11. Thomas, G., Jamaludheen, J., Sibi, L., & Maneesh, P. (2015, April). A novel mathematical model for group communication with trusted key generation and distribution using Shamir's secret key and USB security. In *2015 International Conference on Communications and Signal Processing (ICCSP)* (pp. 0435–0438). IEEE.
12. Shankar, K., Elhoseny, M., Chelvi, E. D., Lakshmanaprabu, S. K., & Wu, W. (2018). An efficient optimal key based chaos function for medical image security. *IEEE Access, 6*, 77145–77154.
13. Elsayed, W., Elhoseny, M., Sabbeh, S., & Riad, A. (In Press). Self-maintenance model for wireless sensor networks. *Computers and Electrical Engineering*. Available online December 2017.
14. Anbarasi, L. J., Mala, G. A., & Narendra, M. (2015). DNA based multi-secret image sharing. *Procedia Computer Science, 46*, 1794–1801.
15. Shankar, K., & Eswaran, P. (2016, January). A new k out of n secret image sharing scheme in visual cryptography. In *2016 10th International Conference on Intelligent Systems and Control (ISCO)* (pp. 1–6). IEEE.
16. Bhadravati, S., Atrey, P. K., & Khabbazian, M. (2015). Scalable secret sharing of compressed multimedia. *Journal of Information Security and Applications, 23*, 8–27.
17. El-Shorbagy, M. A., Elhoseny, M., Hassanien, A. E., & Ahmed, S. H. A novel PSO algorithm for dynamic wireless sensor network multiobjective optimization problem. *Transactions on Emerging Telecommunications Technologies*.

18. Do, Q., Martini, B., & Choo, K. K. R. (2018). The role of the adversary model in applied security research. *Computers & Security*.

19. Shankar, K., Elhoseny, M., Kumar, R. S., Lakshmanaprabu, S. K., & Yuan, X. (2018). Secret image sharing scheme with encrypted shadow images using optimal homomorphic encryption technique. *Journal of Ambient Intelligence and Humanized Computing, 1–13*.

20. Shankar, K., & Lakshmanaprabu, S. K. (2018). Optimal key based homomorphic encryption for color image security aid of ant lion optimization algorithm. *International Journal of Engineering & Technology, 7*(9), 22–27.

21. Mirjalili, S., Mirjalili, S. M., & Lewis, A. (2014). Grey wolf optimizer. *Advances in Engineering Software, 69*, 46–61.

22. Shankar, K., & Eswaran, P. (2016). RGB-based secure share creation in visual cryptography using optimal elliptic curve cryptography technique. *Journal of Circuits, Systems and Computers, 25*(11), 1650138.

23. Chum, C. S., & Zhang, X. (2013). Hash function-based secret sharing scheme designs. *Security and Communication Networks, 6*(5), 584–592.

Chapter 10
Optimal Stream Encryption for Multiple Shares of Images by Improved Cuckoo Search Model

Abstract Security of the media information such as image and video is one of the fundamental prerequisites for broadcasting communications and computer systems. Most of image security system, for famous systems, similar to the Internet, is not reasonable for wireless sensor systems, requesting legitimate examination around there. In the proposed investigation, the security of Digital Images (DI) is upgraded by utilizing a Lightweight encryption algorithm. To ensure security, the images selected for security investigation was shared by a number of copies. With the help of the presented share creation model, for example, Chinese Remainder Theorem (CRT), the image was encoded into a number of shares and the created shares were highly secured by the proposed Stream Encryption (SE) algorithm. The metaheuristic algorithm was introduced by passing the selection of optimal public and private keys to encrypt as well as decrypt the image in secure transmission. With the help of Improved Cuckoo Search Algorithm (ICSA), the optimal keys were selected with fitness function as maximum throughput. The proposed SE-based ICSA algorithm consumed the least time in creating key value to decrypt the image in WSN security model. The simulation result exhibited that the SE-based ICSA algorithm enhanced the exactness of DI security for all input images (Lena, Barbara, baboon, house, and airplane) when compared with existing algorithms.

Keywords Image security · Share creation · Lightweight SE · Key optimization · ICSA · Throughput · PSNR · WSN

10.1 Introduction

Visual Cryptography is a unique encryption system to hide data in images so that it very well is decrypted from human vision if the right key image is utilized in WSN security [1]. It manages the system utilized to revamp the data among reasonable and immeasurable structures by making use of encryption and decryption strategies under the intensity of the keys. It gives substance security and access control [2]. All shares are imprinted on alternate clarity and decryption is finished by a strategy for superimposing the shares. When all the "n" shares are exactly superimposed, it

© Springer Nature Switzerland AG 2019

K. Shankar and M. Elhoseny, *Secure Image Transmission in Wireless Sensor Network (WSN) Applications*, Lecture Notes in Electrical Engineering 564,
https://doi.org/10.1007/978-3-030-20816-5_10

enables the underlying image to end obviously [3]. The security strategies required are symmetric and asymmetric algorithms, hybrid techniques, pre-dispersion key calculation and intermediate node-based techniques [4]. A visual information pixel synchronization and blunder dispersion system empower the encryption of visual data with great superiority [5]. The threshold strategy, associated with the visual cryptography, makes the application less demanding which in turn reduces the difficulty [6]. The secret image must be shared among the members. The image is partitioned into 'n' shares so that if 'm' transparencies (shares) are put together, the image gets noticed in sensor nodes. Numerous WSN applications will have security prerequisites. Sensor nodes might be conveyed in expansive and difficult to-get to regions, where the remote channel may be gotten to by unauthorized individuals [7]. At the point when there are fewer transparencies, it is undetectable [8]. The human eyes can decrypt secrets effectively where decryption remains troublesome for a computer. In common secret sharing plans, the secret is either numbers or message, and in VCSs, the secret is an image [9]. These systems are split into a number of shares and at decryption side, all or a portion of the shares are covered with one another to uncover the secret image for networking model [10].

The benefit of utilizing stream cipher is that the execution speed is higher when compared with block cipher and it has lower equipment unpredictability [11]. Unlike block cipher, the stream cipher do not create a similar cipher text notwithstanding for redundant blocks of plain content, since the keys are changed always for all the plain content [12]. Keystream is a gathering of characters indicating the keys for encryption. Depending upon the distribution of characters in plain image speaking to the encoded binary image, a character code tree is shaped to produce the code for every pixel value [12]. The original image is encoded by practice stream cipher approach and in this way, the encrypted image is implanted with practical lossless data hiding strategy. Inside the beneficiary perspective, reversible data hiding algorithmic guideline is connected with the encrypted image [13]. Nevertheless inalienable issues when endeavoring to guarantee classification, the transmission stream may likewise be liable to uprightness assaults of WSN security. Finally, validation is likewise required for some applications, so as to assure that recovered data originates from substantial source nodes in senor networking [14].

10.2   Recent Literature

Author/Year/Ref	Technique	Description	Performance parameters
Shankar et al. (2018) [15]	Homomorphic encryption and adaptive whale optimization (AWO)	A key-based homomorphic encryption (MHE) with (AWO) algorithm was proposed in this study. The fitness function was considered for improvement of PSNR among plain and cipher images. The first image was changed into original image which was transformed into blocks and afterward adjusted with the help of encryption process	PSNR, CC, entropy, key sensitivity, and MSE
Li et al. (2018) [16]	Visual secret sharing (VSS)	A staggered and significant-shared VSS plot depends on generalized stochastic grids. The recouped secret image did not contain data about overlay images amid recuperation. The diverse areas seem distinctive normal light transmissions bringing about staggered overlay image sharing	PSNR
Liu and Chang (2018) [17]	Visual cryptography	Turtle shell (TS)-based VSS scheme shares the secret data in two ways. First, a TS reference matrix is established and then secret data is hidden in a cover image with the help of TS reference matrix	PSNR, detection ratio, visual quality
Shankar et al. (2018) [18]	HE, discrete wavelet transform (DWT), oppositional based harmony search (OHS)	Numerous shadows were made and encryption and decryption was performed for every shadow. The encrypted shadow can be recuperated by just picking some subsets of these 'n' shadows that makes it straightforward and stack over one another. To enhance shadow security, each shadow was encoded and decrypted by making use of HE strategy	PSNR, entropy, MSE, MAE, CC

(continued)

(continued)

Author/Year/Ref	Technique	Description	Performance parameters
Al-Khalid et al. (2017) [19]	VSS model	To encrypt halftone color images by producing two shares, arbitrary and key shares which are a similar size as the secret color image. Shares are produced dependent on a private key. At the accepting side, the secret color image is uncovered by stacking the two shares and abusing the human vision framework. An improved type of the proposed strategy by adjusting the encryption method used to create the random and the key shares	PSNR
Shankar and Eswaran (2015) [7]	ECC and share generation	The first image, in fact, was acknowledged by covering the whole shares straightforwardly. The unique method is used to create different shares which were exposed to encryption and decryption processes through elliptic bend cryptography procedure	PSNR, MSE, CC

10.3 Background of Stream Encryption in Image Security

Usually, SEAs are utilized for image encryption. Stream ciphers are constructed utilizing a pseudo-arbitrary key succession after which the arrangement is joined with the initial text through exclusive operator. By and large, stream encryption frameworks exhibit high or reasonable execution in terms of speed as well as error probability of information transmission. In this chapter, the basic and lightweight SEA are utilized for multimedia applications, for example, image. Further, different factual tests are performed so as to guarantee the security of the algorithm. Security in remote sensor systems might be difficult to accomplish because of numerous components [20]. The first of them is the asset compelled nature of sensor nodes.

10.4 Proposed System: Enhancing Image Security by Share Generation

The proposed work intends to improve image security by means of share creation and lightweight cipher-based implementation. The image taken for security analysis is shared by number of copies. With the assistance of the presented share creation model (CRT), the image was encrypted into a number of shares and the created shares were sent by means of ensured communication independently towards the cloud. Furthermore, at the decryption side, all or some of the shares were covered with one another to uncover the secret image. In addition, some critical questions emerge while applying encryption plans to WSNs like, how the keys are created or scattered. In the process of encryption and decryption, the optimal keys need to be found which enhances the image security. For the image transmission model, AODV routing protocol used, here Mobile Adhoc Network (MANET) structure used.

10.4.1 Share Generation of Image in WSN

In order to improve the image security level, it is possible to generate a number of shares, Chinese Remainder Theorem (CRT), an innovative offer share model is utilized to achieve this. CRT is issued to work out a set of synchronous harmoniousness conditions which can be delineated as pursues. The generated shares in one sensor node to another sensor added to the system or restored for guaranteeing energetic security. Although numerous remote wireless sensor systems will require some dimension of dependability for the transmission of visual information, scalar and control information may stream in a temperamental route in an expansive arrangement of observing and control scenarios [21].

Let us assume co-prime positive integers as q_1, q_2, q_3, ... q_k and the non-negative integers as c_1, c_2, c_3, ... c_k. There exists exactly one solution i.e. $s \in [q_1, q_2, q_3, ... q_k)$ for simultaneous congruence evaluation, which is performed by the following equation.

$$
\begin{aligned}
s &\equiv c_1 (\text{mod } q_1) \\
s &\equiv c_2 (\text{mod } q_2) \\
&\vdots \\
s &\equiv c_k (\text{mod } q_k)
\end{aligned}
\tag{10.1}
$$

The desired solution i.e. number of share creation can be performed by executing the following CRT steps.

<u>Step 1:</u> Allocate $Q = \Pi_{i=1}^{k} q_i$ and assume $m_i = Q/q_i$, where the value of i ranges from [1, k]

<u>Step 2:</u> Assume $n_i \equiv m_i^{-1} (\text{mod } q_i)$ for $1 \leq i \leq k$

<u>Step 3:</u> Let $s \equiv (c_1m_1n_1 + c_2m_2n_2 + c_3m_3n_3 + \cdots + c_km_kn_k) \bmod Q$
<u>Step 4:</u> In the fourth step, $s \equiv c_i(\bmod q_i)$ was chosen as the unique solution of the algorithm for $1 \leq i \leq k$. An example stated as s = 39 (mod $2 \cdot 5 \cdot 7$) which is the unique solution for following congruence equations.

$$s \equiv 1(\bmod 2), \ s \equiv 4(\bmod 5), \ s \equiv 4(\bmod 7) \tag{10.2}$$

Based on this CRT technique, the numbers of shares were generated; this enhances the image security. Further, a security model SEA was presented to secure the generated shares in WSN.

10.4.2 Shares Security: Stream Encryption Algorithm

In this algorithm, the primary text was partitioned into various segments and each segment was encrypted by SEA. In any segment, the encryption algorithm utilized a partition secret key. The secret key of the encryption was ensured by the block cipher (for example, AES). The diagrammatic representation of the SE algorithm is represented in Fig. 10.1. By assuming F be a function which can be characterized as

$$f(key_i, P) = ((((P \times key_1) \oplus key_2) + key_3) \oplus key_4) \tag{10.3}$$

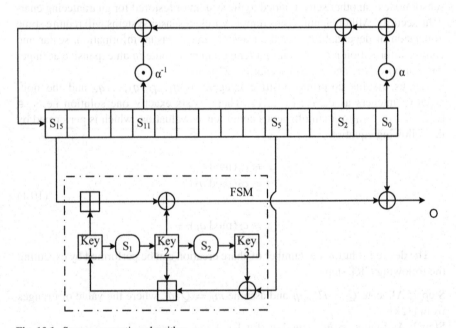

Fig. 10.1 Stream encryption algorithm

where key_i is the 128-bit key and $key_i = key_1 key_2 key_3 key_4$, p indicates a 32-bit string, the symbol $\oplus$ represents the bit-wise exclusive-or, $+$ and $\times$ are mod 232 addition as well as multiplication. For encrypting every 32 bits of the plain text, this algorithm had the following steps.

(i) A 128-bit key sequence was created by the block algorithm AES and was considered to be $key_i = key_1 key_2 key_3 key_4$ for 32-bit key_i. The function in Eq. (10.3) is expressed as in three stages and at each stage, the P value was replaced by the corresponding cipher and plain text.

$$\text{Stage 1: } a_i = f(key_i, Cipher_{i-1} \oplus Plain_{i-1}) \tag{10.4}$$

$$\text{Stage 2: } b_i = f(key_i, a_i \oplus Plain_{i-2}) \tag{10.5}$$

$$\text{Stage 3: } c_i = f(key_i, b_i \oplus Cipher_{i-2}) \tag{10.6}$$

(ii) In view of the accompanying equation, 32 bits of the ciphertext were attained:

$$Cipher_i = Plain_i \oplus c_i \tag{10.7}$$

where $Plain_i$ is the value equal to 32 bits of the plain text and finally the values are summarized as in Eq. (10.8).

$$Cipher_i = Plain_i \oplus f(key_i, f(key_i, f(key_i, Cipher_{i-1} \oplus Plain_{i-1})$$
$$\oplus Plain_{i-2}) \oplus Cipher_{i-2}) \tag{10.8}$$

Decryption
The decryption procedure is similar to encryption procedure with few differences i.e., the locations of $Cipher_i$ and $Plain_i$ (10.7) with public key, are interchanged as below.

$$Plain_i = Cipher_i \oplus c_i \tag{10.9}$$

It ought to be referred that c_i value in the decryption technique was acquired as per the encryption methodology as utilized in the past original as well as encoded texts.

10.4.3 Optimal Public and Private Key Selection in SEA

From a number of keys produced, the optimal one was chosen (it acts as a public key for encrypting and private key for decrypting of the image) for the end users. With the intention of choosing an optimal key, an enhancement algorithm was proposed which advances the value as minimum or maximum depending on the target of the current study aims. Here, the ICS algorithm was used which selects the optimal keys to secure the image considered. The complexity nature of both encryption and decryption calculations is critical for conventional and particular encryption alike. These cryptographic methodologies, message confirmation codes, symmetric keys, and open key encryption for image security in secure transmission (WSN) process.

Objective function

When the greatest security is achieved for a given key size, it can be utilized to solve the issue of finding the image security level and for that, ICS is implemented. The optimal key was picked depending on the greatest throughput which is expressed in condition (10.10)

$$OF_{security} = \text{Max\_throughput} \tag{10.10}$$

10.4.4 Improved CS Algorithm

CS is a developmental optimization algorithm which is motivated by cuckoo bird types which are nothing but 'Brood parasites' fowls. It never assembles or builds its own nest but lays its eggs in the nest of other host birds' nests. Some host birds can connect straightforwardly with the meddling cuckoo. On the off chance, when the host bird recognizes the eggs that are not theirs, it either discard those eggs from its nest or essentially empty its nest and construct new one. The General CS behavior is shown in Fig. 10.2.

The CS enhancement algorithm is essentially developed based on the accompanying three standards:

- Each cuckoo chooses a home haphazardly and lays one egg in it.
- The best homes with high caliber of eggs get extended to the people who arrives.
- For a settled number of homes, a host cuckoo can find an outside egg with probability $\in [0, 1]$. In this situation, the host cuckoo can either discard the egg or forsake the home and assemble another one elsewhere.

From the point of execution, the representation pursued is that each egg in a home indicates an answer, and each cuckoo can lay only a single egg (in this way symbolizing one solution). It is possible to safely make no regard among an egg, a home or a cuckoo. The point is to utilize the new and possibly a better solution (cuckoo egg) to replace a bad solution in the home.

Levy Flight: Contrasted with ordinary random walks, Lévy flights are increasingly productive in investigating large-scale search zones. This is essentially due to Lévy flights' fluctuations that increase quicker than that of the ordinary random walk. Lévy flights can diminish the quantity of enhancement algorithm iterations by four orders contrasted with normal algorithm. CS is an efficient algorithm because it maintains a

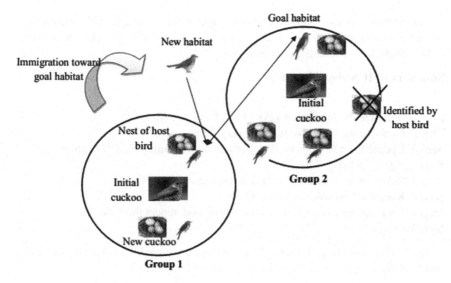

Fig. 10.2 Searching behavior of cuckoo bird

good balance between local and the global random walk. The local as well as global random walks are described in Eqs. (10.11) and (10.12) along with Lévy distribution (to find the step size of the random walk).

$$R_i^{new} = R_i^t + \eta S \otimes F(SP_a - \varepsilon) \otimes (R_j^t - R_k^t) \tag{10.11}$$

$$R_i^{new} = R_i^t + \eta L(S, \delta) \tag{10.12}$$

Variable Declaration: R_i^t, R_k^t symbolize the current positions selected by random permutation. The term η indicates the positive step size scaling factor, R_i^{new} denotes the next position and S denotes the step size. SP_a represents the switching parameter and $L(S, \delta)$ denotes the Lévy distribution.

Improvements in the CS Algorithm

Normally, the switching parameter between local and global random walks in CS is kept at constant. The proposed ICS algorithm enthusiastically altered the value of the switching parameter using Eq. (10.13).

$$SP_{ai} = (SP_{a\,MAX}) * Exp(I_i/I_T) \tag{10.13}$$

The expansion of Eq. (10.13) is SP_{ai}, $(SP_{a\,MAX})$ which denotes the current and maximum values of switching parameters. $Exp(I_i/I_T)$ is the exponential function of current and total number of iterations.

Termination process: Until achieving the maximum throughput with optimal public and private keys, the process was repeated. The proposed ICS worked on the basis of levy flight and switching parameter only.

Summary of ICS algorithm steps

Step 1: Initialize the number of keys $(Key_1 Key_2, \ldots Key_n)$
Step 2: Evaluate the objective function (Eq. 7.5)
Step 3: Update the new keys by levy flight and switching parameter values
Step 4: Again find (Eq. 7.5)
Step 5: Select $\{new\ solution \quad if\,(\text{Fitness (new)} >\text{fitness})$
Step 6: Keep the best solution and rank it
Step 7: Until the maximum throughput is achieved, repeat the process
Step 8: End

Based on the above procedure, the image was encrypted with optimal public key and then the image was decrypted using the private optimal key.

10.5 Result and Analysis

Enhanced security scheme results were analyzed in this section by security measures such as PSNR, Entropy, NPCR and maximum throughput values. It was implemented in MATLAB 2016 with an i5 processor and 4GB RAM.

Table 10.1 explains the security measures for the image security analyses of the proposed SE-ICSA model. For the image security analyses, the images considered were Lena, Barbara, baboon, house, and airplane. The investigated measures such as PSNR, entropy, NPCR, and throughput are illustrated in the table; this deliberates the level of image security.

The image representation of each step in the security model is illustrated in Fig. 10.3. Here, five images were considered to analyze the security level. Using the proposed CRT share generation model, two shares of each input image was created and is shown in figure (c). From the shared copies, the original image can be attained by decrypting it using SE algorithm.

Table 10.1 Security measures for the proposed SE-ICSA

Images	PSNR (dB)	Entropy	NPCR (%)	Throughput
Lena	63.225	7.99	99.48	26.45
Barbara	59.22	7.68	99.85	29.44
Baboon	60.12	7.92	98.45	23.22
House	58.44	7.85	96.45	35.44
Airplane	58.78	7.99	99.56	32.22

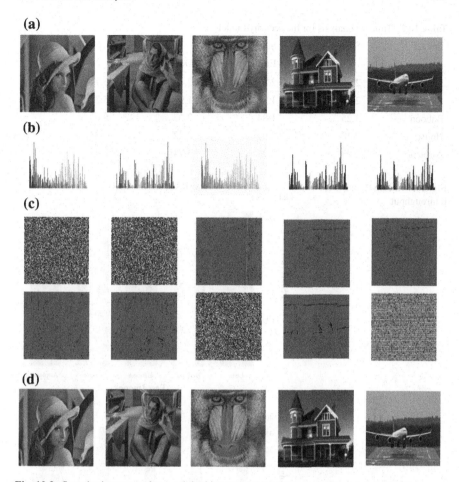

Fig. 10.3 Security image results: **a** original images, **b** histogram, **c** shares and **d** stacked image

Table 10.2 describes the execution time analyses of the security model for five input images. For example, the Lena image took 2.85 s for share creation and in case of encryption, it was 2.22 s and for decryption, 2.45 s was taken. The overall computational time was 5.74 s. Likewise, the time taken for other four images are explained in Table 10.2.

Figure 10.4 explains the PSNR and throughput values for the key optimization model i.e. SE-ICSA. The performance of the proposed model was compared with SE-CSA and SE algorithms. From the bar graph analysis, the maximum PSNR and throughput were attained in the proposed model SE-ICSA for all the five images i.e., Lena, Barbara, baboon, house, and airplane.

The measures such as NPCR and entropy were analyzed and compared with key optimization model and is detailed in Table 10.3. For the proposed model, the entropy values for every image were as follows, it was 7.45, 7.66, 7.55, 6.57 and 7.35 for

Table 10.2 Time (s) analysis for the security model

Images	Share creation	Security		Computational time
		Encryption	Decryption	
Lena	2.85	2.22	2.45	5.74
Barbara	2.35	1.88	1.99	5.2
Baboon	3.25	1.84	2.07	8.09
House	2.48	2.99	3.22	7.08
Airplane	2.88	2.78	3.74	7.48

Fig. 10.4 a PSNR and
b throughput

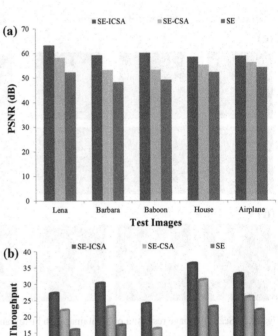

Lena, Barbara, Baboon, House and Airplane images respectively. The NPCR value of Lena image was 92.22 whereas for Barbara, it was 93.22. For Baboon, it was 94.11 and for house, it was 89.45 whereas in case of Airplane, it was 88.22. Compared to existing key optimization models, the proposed one achieved high entropy and NPCR value.

Figure 10.5 explains the attack-applied PSNR results. After applying salt and pepper noise, the noise attacked image was extracted from the digital image. For

Table 10.3 Comparative analyses of NPCR and entropy

Images	Entropy		NPCR (%)	
	SE-CSA	SE	SE-CSA	SE
Lena	7.45	7.22	92.22	82.14
Barbara	7.66	6.77	93.22	83.22
Baboon	7.55	6.88	94.11	79.55
House	6.57	6.47	89.45	72.11
Airplane	7.35	7.22	88.22	78.55

Fig. 10.5 Attack applied PSNR results

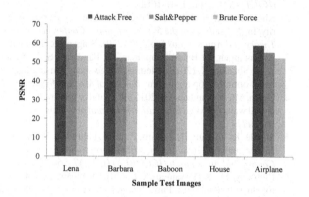

example, for the house image, the PSNR of brute force was 58 dB, salt and pepper was 59 dB and when attack free, the value was 60 dB. Similarly, the results for other images are depicted in the graph.

10.6 Conclusion

In this chapter, the security level of five different images such as Lena, Barbara, baboon, house, and airplane were examined. This chapter implemented the share creation model (CRT) to secure images by generating two numbers of shares. Then the created shares were highly secured using a lightweight SE algorithm. In SE algorithm, the optimal key is to be found using an innovative metaheuristic optimization i.e. ICSA where the most optimal keys are selected for the encryption and decryption processes. Security systems might be basic in WSN structure. Recent works have concentrated on imaginative components to give distinctive levels of security relying upon the accessible assets of sensor networks. The key selection was done based on the maximum throughput by Levy flight and switching parameter normalization. The simulation results concluded that the proposed SE-ICSA accomplished the desired PSNR, entropy, and NPCR values with minimum computational time. For future research on image security, a detailed explanation of different encryption algorithms will be investigated with various metrics using innovative data along with the hybrid optimization approach. This will be applicable for various purposes in cloud image security under different threatening circumstances.

References

1. Patel, T., & Srivastava, R. (2016, August). A new technique for color share generation using visual cryptography. In *International Conference on Inventive Computation Technologies (ICICT)* (Vol. 2, pp. 1–4). IEEE.
2. Khokhar, P., & Jena, D. (2017). Color image visual cryptography scheme with enhanced security. In *Proceedings of the 5th International Conference on Frontiers in Intelligent Computing: Theory and Applications* (pp. 267–279). Singapore: Springer.
3. Avudaiappan, T., Balasubramanian, R., Pandiyan, S. S., Saravanan, M., Lakshmanaprabu, S. K., & Shankar, K. (2018). Medical image security using dual encryption with the oppositional based optimization algorithm. *Journal of Medical Systems, 42*(11), 208.
4. Gupta, M., & Chauhan, D. (2015). A visual cryptographic scheme to secure image shares using digital watermarking. *International Journal of Computer Science and Information Technologies (IJCSIT)*.
5. Shinde, K. V., Kaur, H., & Patil, P. (2015, February). Enhance security for spontaneous wireless ad hoc network creation. In *2015 International Conference on Computing Communication Control and Automation (ICCUBEA)* (pp. 247–250). IEEE.
6. Elhoseny, M., & Hassanien, A. E. (2019). Dynamic wireless sensor networks: New directions for smart technologies. Published in *Studies in Systems, Decision and Control*. Springer.
7. Shankar, K., & Eswaran, P. (2015). Sharing a secret image with encapsulated shares in visual cryptography. *Procedia Computer Science, 70*, 462–468.
8. Begum, A. A. S., & Nirmala, S. (2018). Secure visual cryptography for medical image using modified cuckoo search. *Multimedia Tools and Applications*, 1–20.
9. Wadi, S. M., & Zainal, N. (2017). Enhanced hybrid image security algorithms for high definition images in multiple applications. *Multidimensional Systems and Signal Processing*, 1–24.
10. Kita, N., & Miyata, K. (2018). Magic sheets: Visual cryptography with common shares. *Computational Visual Media, 4*(2), 185–195.
11. Do, Q., Martini, B., & Choo, K. K. R. (2018). The role of the adversary model in applied security research. *Computers & Security*.
12. Sreelaja, N. K., & Pai, G. V. (2012). Stream cipher for binary image encryption using Ant Colony Optimization based key generation. *Applied Soft Computing, 12*(9), 2879–2895.
13. Aïssa, B., Nadir, D., & Mohamed, R. (2011, July). Image encryption using stream cipher algorithm with nonlinear filtering function. In *2011 International Conference on High Performance Computing and Simulation (HPCS)* (pp. 830–835). IEEE.
14. El-Shorbagy, M. A., Elhoseny, M., Hassanien, A. E., & Ahmed, S. H. (2018). A novel PSO algorithm for dynamic wireless sensor network multiobjective optimization problem. *Transactions on Emerging Telecommunications Technologies*, e3523.
15. Shankar, K., Lakshmanaprabu, S. K., Gupta, D., Khanna, A., & de Albuquerque, V. H. C. (2018). Adaptive optimal multi key based encryption for digital image security. *Concurrency and Computation: Practice and Experience*, e5122.
16. Li, Y., Xiong, C., Han, X., Du, H., & He, F. (2018). Internet-scale secret sharing algorithm with multimedia applications. *Multimedia Tools and Applications*, 1–6.
17. Liu, Y., & Chang, C. C. (2018). A turtle shell-based visual secret sharing scheme with reversibility and authentication. *Multimedia Tools and Applications*, 1–16.
18. Shankar, K., Elhoseny, M., Kumar, R. S., Lakshmanaprabu, S. K., & Yuan, X. (2018). Secret image sharing scheme with encrypted shadow images using optimal homomorphic encryption technique. *Journal of Ambient Intelligence and Humanized Computing*, 1–13.

19. Al-Khalid, R. I., Al-Dallah, R. A., Al-Anani, A. M., Barham, R. M., & Hajir, S. I. (2017). A secure visual cryptography scheme using private key with invariant share sizes. *Journal of Software Engineering and Applications, 10*(1), 1–10.
20. Elhoseny, M., Farouk, A., Zhou, N., Wang, M.-M., Abdalla, S., & Batle, J. (2017). Dynamic multi-hop clustering in a wireless sensor network: Performance improvement. *Wireless Personal Communications, Springer US, 95*(4), 3733–3753.
21. Elhoseny, M., Tharwat, A., Yuan, X., & Hassanien, A. E. (2018). Optimizing K-coverage of mobile WSNs. *Expert Systems with Applications, 92,* 142–153.

Printed in the United States
By Bookmasters